KT-486-560

DAMPNESS IN BUILDINGS

DAMPNESS IN BUILDINGS

Alan Oliver

Second Edition revised by

James Douglas and J. Stewart Stirling
Department of Building Engineering & Surveying
Heriot-Watt University, Edinburgh

b

Blackwell
Science

Blackwell Science Ltd
Editorial Offices:
Osney Mead, Oxford OX2 0EL
25 John Street, London WC1N 2BL
23 Ainslie Place, Edinburgh EH3 6AJ
238 Main Street, Cambridge
    Massachusetts 02142, USA
54 University Street, Carlton
    Victoria 3053, Australia

Other Editorial Offices:
Arnette Blackwell SA
    224, Boulevard Saint Germain
    75007 Paris, France

Blackwell Wissenschafts-Verlag GmbH
    Kurfürstendamm 57
    10707 Berlin, Germany

    Zehetnergasse 6
    A-1140 Wien
    Austria

First edition published by BSP Professional Books
    1988
Reissued in paperback 1991
Reprinted 1993
Second edition published by Blackwell Science Ltd
    1997

Set in 10 point Times
by DP Photosetting, Aylesbury, Bucks
Printed and bound in Great Britain
by Hartnolls Ltd, Bodmin, Cornwall

The Blackwell Science logo is a trade mark
of Blackwell Science Ltd, registered at the
United Kingdom Trade Marks Registry

ACKNOWLEDGEMENTS
Material reproduced from Building Research
Establishment Digests is Crown copyright, and is
reproduced by permission of the Controller of HM
Stationery Office.
    Extracts from British Standards are reproduced by
permission of BSI. Complete copies can be obtained
from them at Linford Wood, Milton Keynes, MK14
6LE.

DISTRIBUTORS

Marston Book Services Ltd
PO Box 269
Abingdon
Oxon OX14 4YN
(Orders: Tel: 01235 465500
                Fax: 01235 465555)

USA
Blackwell Science, Inc.
238 Main Street
Cambridge, MA 02142
(Orders: Tel: 800 215-1000
                     617 876-7000
                Fax: 617 492-5263)

Canada
Copp Clark Professional
200 Adelaide Street, West, 3rd Floor
Toronto, Ontario M5H 1W7
(Orders: Tel: 416 597-1616
                     800 815-9417
                Fax: 416 597-1617)

Australia
Blackwell Science Pty Ltd
54 University Street
Carlton, Victoria 3053
(Orders: Tel: 03 9347 0300
                Fax: 03 9347 5001)

A catalogue record for this title is available from the
British Library

ISBN 0–632–04085–8

Library of Congress
Cataloging-in-Publication Data

Oliver, Alan
    Dampness in buildings/Alan Oliver.—2nd ed./rev.
by James Douglas and Stewart Stirling.
        p.   cm.
    Includes bibliographical references and index.
    ISBN 0–632–04085–8
    1. Dampness in buildings.   I. Douglas, James.
II. Stirling, Stewart.   III. Title.
TH9031.045   1996
693'.892—dc20   96-20739
                                                                              CIP

Contents

Preface to First Edition

One of the major functions of a building is to exclude or be resistant to excessive moisture, and yet dampness is one of the most common causes of building failure. Dampness in buildings can originate in many ways and can cause a variety of effects on building materials.

In identifying the causes and effects of dampness, many aspects of science and technology must be taken into account – physics, chemistry, biology, as well as building science and technology. In reality the scope of the subject is vast, and in attempting to prepare a useful review of the topic the difficulties are as much what to leave out as what to include. It is to be hoped that the material included is relevant and helpful to the reader, and will provide a useful source of reference when unfamiliar problems arise.

The emphasis throughout has been to assist all who are involved in restoration, rehabilitation and repair of buildings, be they architects, surveyors, engineers, or contractors, as well as keen and practical property owners. Most books on building materials and construction concentrate on new methods and techniques, appropriate to new construction, but the objective of this book is to strike a balance between new and obsolescent methods and materials.

Some of the material is derived from lectures that I have given, over many years, to students attending special short courses at Buckinghamshire College of Higher Education on 'Remedial Treatment of Timber and Dampness in Buildings', as well as from lectures to many groups of surveyors and architects in various parts of the country. These topics have been expanded to include a review of building materials and building construction and their reactions to water in various forms.

I would like to record the great help that I have received from many colleagues and associates in discussion and comment over the years on techniques and basic principles. In particular, I must record special thanks to Dr Colin Kyte, Graham Coleman, David Hyde, Dr Marius Perkins, Geoffrey Sharpe, the late Dr Ian Steele, Mike Hammett, and

Ken Fearnley, and thanks are due to John Bricknell and Barry Matthews for the drawings.

Finally, I would like to dedicate the book to my family: to my wife Judith for her invaluable support during the time of preparation of the book, and to my two sons, James and Stewart, who in their careers are also involved in the efficient and commercial use of buildings in the United Kingdom.

Alan C. Oliver
East Witheridge,
Knotty Green,
Beaconsfield,
Buckinghamshire

1987

Preface to Second Edition

This edition of Alan Oliver's book has been compiled by two chartered building surveyors. The opportunity has been taken to include additional material as well as make some necessary revisions. Another new feature is the list of references contained in the bibliography at the end of the book, which includes a comprehensive list of relevant BRE publications. Important technical terms requiring elaboration are included in a glossary at the end of the book.

The structure of this book is similar to that of the previous edition. The first four chapters discuss the primary sources of moisture, and describe the relevant properties of common building materials that are used in the structure and external fabric of buildings, with some reference to their reactions to dampness. The Introduction, however, is expanded and updated, and is now Chapter 1. It considers the health implications of dampness as well as its effects on buildings. Chapters 2–4 consider moisture in common building materials. Chapter 5 describes the design and construction of the main building elements in relation to moisture exclusion. Chapter 6 deals with condensation, Chapter 7 rising damp, Chapter 8 (a new chapter) penetrating damp, and Chapter 9 considers techniques of diagnosis and therapy. In addition, several new appendices are included in this edition.

Throughout the book, emphasis is placed on existing buildings, which may well be constructed with materials that are now obsolete (e.g. stucco), and with outdated methods (e.g. solid masonry walling). Consideration is given to dampness in modern construction as well, and this is primarily covered in the chapter on penetrating damp. The introduction of new construction techniques to such buildings can often give rise to unforeseen problems, such as incompatibility between materials.

There are now a variety of hand-held moisture meters available for investigating dampness. Still, great care needs to be taken in interpreting the readings from such meters, especially when assessing dampness in masonry materials. Surveyors in particular need to be aware of the use and limitations of those meters, and these issues are covered in the final chapter. In addition, Appendix M summarises some recently published useful guidance for surveyors on the measurement of moisture in masonry.

The aim therefore is to provide an updated, practical assessment of the problem of dampness from a practitioner's perspective. The revisers have built on the sound foundation provided by the first author. A word of caution, however, needs to be stressed. This book cannot claim to cover all possible dampness scenarios. The range of different types of building and their dampness problems is too varied for that. Hopefully, though, the references indicated in this book will provide useful guidance on where further information can be obtained. Naturally, the revisers cannot accept any responsibility for any errors resulting from following the general guidance within this book.

Some of the additional material is based on teaching that the revisers have given to building surveying and other undergraduate construction students at Heriot-Watt University on the subjects of building investigation and maintenance. Another source has been the feedback on the first edition. It is hoped that the perspective taken will enhance the practical emphasis of this edition.

Although the principles of dampness in buildings are well known, it is still a major problem for property occupiers and owners alike. Changes in people's attitudes towards buildings, as well as innovations in construction techniques, have contributed to the continuation of this problem. Greater awareness of the health risks associated with high moisture levels within buildings has made the avoidance of dampness an important objective in modern construction.

We should like to express our gratitude to our colleagues at the Department of Building Engineering & Surveying at Heriot-Watt University, who gave their much-valued advice and guidance on certain parts of this edition. Our especial thanks go to Professor Philip Banfill for his help on the various chemical formulae featured in this book and for revising the appendix on ordinary Portland cement.

We should also like to express our deep appreciation to Ernest G. Gobert, Honorary Life President of Protimeter plc, and Jeff Howell of South Bank University. Their feedback and constructive criticism on the use and limitations of moisture meters in assessing dampness was unhesitatingly offered and gratefully received.

Lastly, it would be remiss of us not to thank Alan Oliver for allowing us the opportunity to revise the first edition of his book. It is not easy for an author to hand over his creation to others, particularly two new writers. The advice and encouragement we received from him was both helpful and freely given. We hope that he will be pleased with the second edition.

Acknowledgements

Diagrams reproduced from the Building Research Establishment, by kind permission of the Controller of HMSO: Crown copyright: Figure 1.1, Table 1.1, Table 2.1, Figure 4.8, Figure 6.8, Figure 7.4 and Figure 8.9.

Figure 8.4 by kind permission of Gower Publishing Ltd.

Table 8.1, Figure 8.12 and Figure 8.15 by kind permission of the Construction Industry Research and Information Association.

Figure 1.7 and Figure 9.7 by kind permission of Butterworth Heinemann Ltd.

Extracts from BS8203: 1987 are reproduced with the permission of BSI. (Complete editions of the standards can be obtained by post from BSI Customer Services, 389 Chiswick High Road, London W4 4AL.)

Figure 3.5 and Figure 8.8 are Crown Copyright and are reproduced by kind permission of the Controller HMSO.

Chapter 1
Introduction

Context

Our relationship to water is ambivalent. Neither humans nor buildings can tolerate too little or too much of it. Water is of course indispensable to life which, as we know it, originated in the sea. After all, it constitutes around 70% of the human body's make-up (Solomon *et al.*, 1996). We lose water quickly from our bodies under low-humidity conditions as a result of perspiration and breathing. This water loss will make us feel cold. In contrast, when we are exposed to too much water we feel saturated, and this too will make us feel cold. If we were deprived of water we would dehydrate and eventually die. Where there is a lack of water in buildings, timbers would shrink, split and crumble; cementitious materials would crack; and rooms would become too stuffy.

The importance of water therefore should not be underestimated. It promotes a variety of positive and adverse effects on buildings and their inhabitants. According to one informative source on its properties and significance:

'Water can act as an acid or a base because it can be a proto donor or proto acceptor. It has great solvent power for ionic compounds and can readily dissolve many non-covalent compounds because of its ability to form hydrogen bonds so readily. It is also colourless, tasteless and odourless.

Water plays a critical role in our lives in many subtle ways. It is a major component in ordinary air. It readily absorbs radiation and therefore its presence has a strong influence on the ability of infra-red energy to penetrate the air. It adheres to most surfaces and in very thin layers does not behave as water. As the molecular layers grow they become more water-like, and this affects the electrical surface conductivity and can cause electrical breakdown and corrosion. Water is also essential for life because all organisms, from the simplest to the most complex, rely on aqueous solutions for metabolism.' (Brundrett, 1990).

Excess water in buildings is called dampness. It can have numerous deleterious effects, as described dramatically by one writer:

'The root of all evil is water. It dissolves buildings. Water is elixir to unwelcome life such as rot and insects. Water, the universal solvent, makes chemical reactions happen every place you don't want them. It consumes wood, erodes masonry, corrodes metals, peels paint, expands destructively when it freezes, and permeates everywhere when it evaporates. It warps, swells, discolours, rusts, loosens, mildews and stinks...' (Brand, 1994).

Dampness is to moisture what noise is to sound: they are both unwanted. There are of course significant differences between these two forms of nuisance. Noise, for example, tends to be intermittent and invisible. Dampness, in contrast, is often visible and persistent.

To use another expression, 'all dampness is water out of place' (Oxley and Gobert, 1994). It is something that ideally should not be in a building in any great quantities, but often is, because of a variety of factors. For example, the activities of the occupants or of outsiders can, through neglect or vandalism respectively, allow moisture to enter a building. It can also occur in a new building if it has been inadequately designed, owing to poor detailing, or has not been allowed to dry out properly (see BRE Digest 163).

Unwanted moisture continues to be the scourge of property occupiers and owners in many European countries. In the UK it has been the subject of a number of government studies: for example, by the Scottish Affairs Committee (House of Commons, 1984a,b). Dampness, however, is not solely a European problem. It also adversely affects buildings in the USA (Christian, 1993) and the Far East (Lim, 1988). Thus although this book will consider the problems of dampness in the context of British buildings, the principles it contains should still be relevant elsewhere as well.

If dampness is allowed to continue unchecked, unhealthy conditions may follow, and the building may deteriorate to the extent that it ultimately becomes uninhabitable. Occasionally, damp conditions arise as a result of no maintenance, and sometimes as a result of incorrect application of new building techniques or inappropriate use of new materials, either during new construction or during remodelling work.

The dampness problem is complex, widespread and of great economic importance. It has stimulated successive UK governments to provide funding through grants and loans to try to improve the situation. Many millions of pounds each year are spent on dampness-related repairs. For example, timber decay is primarily caused by moisture (see Chapter 4). It is estimated that the annual turnover of the timber preservation industry in the UK is in excess of £400 million (Singh, 1994). Many specialist contractors have established diagnosis and treatment services to deal with these problems. Surveyors and architects often receive numerous instructions from clients seeking professional advice on curing dampness. Far too often, however, when these professionals suspect dampness they automatically advise 'further investigation by a specialist'. Such specialists often have a vested interest in selling their own timber preservation and rising damp treatments. In the case of timber preservation, irrigating the masonry is usually specified. As regards suspected rising damp, a chemical retrofit dpc is almost always advised. As will be seen in Chapters 4 and 7, these two therapies are open to question.

The widespread nature of dampness in buildings and the high costs of eradication and cure underline the need for an up-to-date reference text covering its causes and effects and a critical review of the state of the art of remedial treatments. This revised edition is designed to be such a text.

Since the end of World War II, dampness in British buildings has become more acute. The final phase of the non-traditional housing construction boom in the late 1960s incurred incipient problems of condensation and penetrating damp (Chandler, 1992). This was probably one of the main influences in the public's reaction against dampness (Bryant, 1979).

Thus there are several reasons why our tolerance of dampness has declined in recent years:

- ❑ Building users have greater expectations. People in the developed world are less likely to accept living in damp conditions, given the discomfort and known health hazards of doing so. Yet even these days there are still reports of tenants living in excessively damp conditions in the remnants of some discredited system-built housing blocks.
- ❑ Biocontamination is moisture-dependent. Moisture is now a well-known key contributor to the spread of disease. Bacteria, fungi, mites and other micro-organisms thrive in damp conditions.
- ❑ Dampness causes aesthetic as well as physical damage to buildings. It can disfigure internal wall and ceiling finishes, and can stain and soil external wall surfaces. Flooding can cause the most severe form of damp-related damage (BRE Digest 152).

❏ Dampness is a catalyst for a whole variety of building defects. Corrosion of metals, fungal attack of timber, efflorescence of brick and mortar, sulphate attack of masonry materials, and carbonation of concrete are all triggered by moisture.

❏ Unwanted moisture in buildings can adversely affect their use. This is particularly the case with dwellings and in certain types of accommodation. For example, operating theatres and computer suites require damp-free conditions, and the maintenance of humidity levels within strict limits.

Research by the Building Research Establishment (BRE), the UK's premier building research agency, since the 1960s has suggested that dampness accounts for over half of all building defects (Ransom, 1987). In response to these influences the various Building Regulations in the UK have tightened up the requirements regarding the control of moisture in buildings. For example, it is now mandatory to have mechanical extract ventilation in all new kitchens and bathrooms (see section at end of Chapter 6).

The on-going research into housing quality and condition has highlighted the fact that dampness is still a problem in many British dwellings. The 1991 English House Condition Survey found that over 10% of the total stock of about 19.7 million dwellings was affected by condensation, just under 5% had rising damp, and about 15% were troubled by penetrating damp (DoE, 1993). Overall the survey found that nearly 20% of the housing stock (i.e. about 3.9 million) was identified as having some dampness problem. The 1991 Scottish House Condition Survey revealed an even worse situation: in a total stock of some 2.13 million dwellings, condensation was a problem in over 19% of households, and over 13% had experienced some form of dampness (Scottish Homes, 1993).

The nature of water

Water is the least understood and most complex of all familiar substances (Brundrett, 1990). It is a molecule comprising two hydrogen atoms and one oxygen atom: hence its chemical label H_2O. Detailed information on the properties of water and its importance in buildings is fully covered by Addleson & Rice (1991).

Water can be present in three forms: as a solid (ice), as a liquid, and as a gas (water vapour). The humidity of the air is a description of the amount of water vapour present. This vapour is normally invisible, unless the air contains so much water vapour that it becomes saturated. After saturation is reached, additional vapour is condensed as liquid water on any adjacent cold surface (such as window panes, metal frames, or wall tiles). The amount of water vapour present in the air at any time is expressed as a humidity relative to the saturation level (hence the term 'relative humidity' – see 'Measurement of moisture' on page 17).

Studies indicate that a family of five people in a house can produce 5 litres every 24 hours (BS 5250: 1995). If humidity levels are lowered, this water loss will increase. Thus our comfort zone, so far as atmospheric humidity is concerned, is generally recognised as being in the range of 40–60% relative humidity (RH). The objective is not to achieve absolute dryness (which itself would be deleterious and uncomfortable), but a moisture content within the range that is comfortable for the occupant. While excessive dryness is to be avoided, excessive humidity, over 70% RH, will lead to unhealthy conditions for the occupants, mould growth, and deterioration of the fabric, decorations and ultimately the contents of the building. The aim is therefore to keep the internal conditions within the comfort zone and to avoid excesses.

Water vapour is expressed in terms of humidity relative to the maximum vapour content possible in the air at that temperature, as warmer air can contain a higher content of water vapour than cold air. At a relatively high humidity, e.g. 80% RH, if the temperature drops then the dew-point temperature is soon reached and condensation occurs on adjacent cold surfaces. Forms of heating, and methods of cooking, washing and drying clothes can all contribute to the vapour levels of the air in a building, increasing humidity, and thus the probability of condensation.

Effects of dampness on building materials

Physical effects

When water enters a building, or is generated in excessive amounts within it, it can have a variety of effects on construction materials: biological (e.g. fungal attack of timber), chemical (e.g. sulphate attack of renders), and physical (e.g. expansion of brickwork). Moreover, cold wintry conditions can cause penetrating rainwater to freeze, and the expansion of this penetrating water on freezing often causes deterioration of any building materials saturated by it. Deterioration of materials as a result of frost action is caused by the expansion of water (9%) in the pores of interstices of the material.

The amount of distribution of moisture in the material, and the material's porous structure, are both thought to be significant in determining whether a material is more or less resistant to frost damage (see test described in BS 3921: 1995). For this reason, the greatest care has to be taken to select resistant materials if they are to be used in exposed conditions where high degrees of water saturation are to be expected (Fig. 1.1).

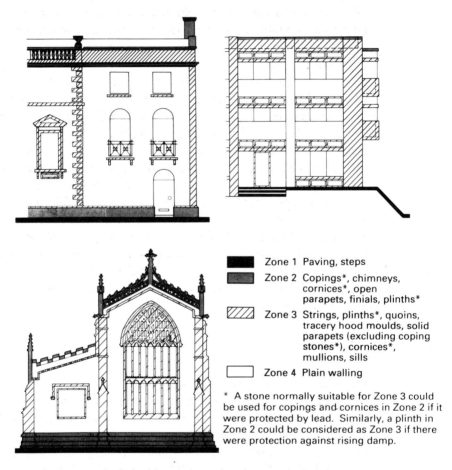

Zone 1 Paving, steps

Zone 2 Copings*, chimneys, cornices*, open parapets, finials, plinths*

Zone 3 Strings, plinths*, quoins, tracery hood moulds, solid parapets (excluding coping stones*), cornices*, mullions, sills

Zone 4 Plain walling

* A stone normally suitable for Zone 3 could be used for copings and cornices in Zone 2 if it were protected by lead. Similarly, a plinth in Zone 2 could be considered as Zone 3 if there were protection against rising damp.

Fig. 1.1 Exposure zones on a building (Ross and Butlin, 1989).

Parapet walls and parts of a building below damp-proof course level are most vulnerable to exposure and frost damage. Gable walls and projections are also vulnerable unless protected from water penetration. High levels of internal insulation can sometimes increase the risk of frost damage to the envelope on the cold side of construction.

Chemical effects

Penetrating water may often contain mineral salts originating from the ground or derived from acid gases such as carbon dioxide (CO_2) and sulphur dioxide (SO_2) in the atmosphere. In addition, penetrating water frequently acts as a solvent for the mineral salts naturally present in most building materials. They are even present in timbers that are impregnated with inorganic salts preservatives. When drying out occurs, these salts come out of solution and accumulate as crystals, or dry salts, which may often be visible as efflorescence. They may also accumulate in the interstices or pores of the building material where, unseen, they may cause erosion, flaking or ultimate deterioration of the 'contaminated' material, because the process of crystallisation often involves swelling, and considerable micro-stresses are generated (i.e. crypto-fluorescence). Some of these salts are hygroscopic and have deliquescent properties, and under high humidities attract moisture, which results in more moisture accumulating in the structure. Such combined hygroscopic moisture is naturally present in many building materials, and becomes an increasingly important factor under high-humidity conditions.

The effects of dampness on exposed building materials and the consequential effects of frost and salts deposition are major contributors to what we call 'weathering' of building materials (see Addleson & Rice, 1991, for a full treatment of this phenomenon). However, other factors implicit in weather exposure, such as sunlight and ultraviolet effects, thermal movement and wind, also play an important part in the ultimate performance of a material.

Biological effects

Biological growths on exterior building materials are a sure indication of dampness. This may be as a result of defective rainwater goods, or constant driving rain or high-humidity conditions constantly wetting exterior surfaces. The effects of these, depending on the location and exposure, are biological growths (Table 1.1). Fungal attack, one of the main biological effects of dampness, is addressed in Chapter 4.

Incompatibility of building materials

Most building materials are inherently inert, even in the presence of some moisture. However, the interaction between different materials in

Table 1.1 Biological growths on exterior building materials (from BRE Digest 370).

Organism	Requirements for		Appearance	Remarks
	Food	Light		
Algae	Mineral salts	Necessary	Green, red or brown powdery deposits	Found on all types of substrate
Lichen	Mineral salts	Necessary	Leathery encrustations: orange, green, grey or black	Found on all types of substrate, especially in rural areas
Mosses	Mineral salts	Necessary	Cushions of green spiky tufts	On surfaces where salts and soil have accumulated: e.g. roofing materials
Liverworts	Mineral salts	Necessary	Leafy, close-growing plants	On surfaces where salts and soil have accumulated: e.g. on stone walls
Moulds	Organic material	Unnecessary	Spots or patches in a variety of colours	Occurs on painted surfaces: e.g. whitewashed walls
Bacteria	Unnecessary	Unnecessary	Not visible to the naked eye	May cause deterioration to some stone and corrosion to metals

juxtaposition can give rise to harmful effects. These arise because of the problem of incompatibility, and moisture is a major catalyst in many such situations.

Many building materials, of course, are highly compatible with each other: plaster and brickwork, for example. Others, however, are not; even similar materials may be incompatible – sandstone and limestone, for example – if they are juxtaposed with each other (Richardson, 1991). There are basically two forms of incompatibility:

❑ Physical incompatibility occurs where different materials have different rates of thermal and moisture movement. Clay bricks and concrete are not very compatible, because the former tend to shrink initially, whereas the latter tend to expand. This can give rise to bowing and cracking of clay brickwork panels in concrete-framed buildings.
❑ Chemical incompatibility occurs where a chemical reaction is triggered between two materials, causing deterioration of one or both. Again, moisture is a key player in this form of incompatibility. Examples of chemical incompatibility are bimetallic or galvanic corrosion, sulphate attack on brickwork, and deplasticisation of polymers in contact with bituminous products.

Incompatibility problems can be avoided by careful selection of juxtaposed materials, using protection or separation barriers, and impermeable barriers against moisture.

In summary, the common deterioration mechanisms in buildings influenced or caused by inadequate moisture control have been identified by Connolly (1993):

❑ hydrolysis;
❑ osmotic pressure;
❑ alkali–silica reactivity;
❑ delayed ettringite formation;
❑ cyclic freeze/thaw degradation;
❑ micro-organism attack;
❑ vapour pressure;
❑ rising damp/salt migration/efflorescence;
❑ corrosion (oxidation);
❑ hygroscopicity;
❑ dissolution;
❑ wetting and drying;
❑ dehydrohalogenation;
❑ plasticizer migration.

Connolly then went on to list the types of deterioration driven or affected by water in building materials:

❑ spalling;
❑ peeling;
❑ delamination;

❑ blistering;
❑ shrinkage, cracking and crazing;
❑ irreversible expansion;
❑ embrittlement;
❑ strength loss;
❑ staining, discoloration;
❑ decay from micro-organisms.

No building material is safe from all these forms of deterioration initiated or aggravated by water. It is not only buildings that can be adversely affected by moisture; their occupants can, too.

Effects of dampness on people

It is reckoned that we spend on average about 90% of our life indoors. Moisture is an important influence on the quality of a building's internal enviroment. There is consistent evidence of a link between damp and mouldy housing and reports of respiratory problems in children (Strachan, 1993). Granted, there are other factors, such as poverty, that can have an influence on ill-health as well (Markus, 1993).

It is well known, however, that moisture in buildings can have at least eight kinds of influence on people (Brundrett, 1990):

❑ warmth, comfort and stress;
❑ dry throats and noses;
❑ eye comfort;
❑ skin comfort;
❑ clothing and fabrics;
❑ electrostatic shocks;
❑ ill-health and allergies;
❑ pollution.

Buildings afford a wide range of niches that provide dead organic material (e.g. dander from animals, mite droppings, skin flakes). They also provide the space for and sources of moisture essential to biological growth. It is thought that this dead organic material and moisture can cause a number of building-related illnesses, owing to their allergenic and toxic effects. Dust, dander and pollen are the main sources of airborne allergens. Typical micro-organisms that thrive inside buildings are bacteria, fungi, moulds, mites and yeasts (Anon, 1987c). Research by environmental and medical scientists has indicated that such organisms can encourage ill-health or lead to disease if the indoor climatic conditions of a building are suitable (Hunt, 1989; Platt *et al.*, 1989; Sundell, 1994).

Mites are minute creatures virtually invisible to the naked eye, and are especially linked to dust from bedding, upholstered furniture, carpets, curtains and other soft furnishings (Singh, 1993). The most important species is the common house dust mite (*Dermatophagoides pteronyssinus*), which is a small (250–350 µm long) relative of the spider. Mites are found in all households, even though each room might be cleaned and dusted regularly. They live in pillows, mattresses, duvets and blankets, by being fed a constant supply of dead human skin flakes. There is a growing body of scientific evidence that suggests that house dust mite droppings (faecal fragments) produce antigens. These compounds are thought to cause respiratory allergy in humans. The two main health problems linked to dust mites are asthma and rhinitis (Brundrett, 1990). Their numbers are heavily influenced by the time of year, being highest in the warmer months – as are levels of bacteria and fungi.

Moreover, a modern building's internal environment is not necessarily more pollution free than the external atmosphere. The concentrations of hazardous substances, such as volatile organic compounds (e.g. benzine from solvents in paints and glues), gases (e.g. carbon monoxide from gas appliances and cigarette smoke), formaldehyde (from pressed wood products, foam insulation and household cleaners) and health-threatening organisms such as bacteria and fungal spores may actually be greater indoors than outside. As a result, occupants of some buildings can be exposed to indoor air pollution. Although the risks to health from such conditions are not fully undertood, a high moisture level in buildings is a vital ingredient in the development of micro-organisms and mites (Singh, 1993). The link between exposure to these organisms and illnesses such as asthma, especially in young children, has still to be established conclusively. Still, some scientists believe there is enough evidence to demand measures to reduce mite numbers in dwellings: for example, by keeping humidity levels as low as is comfortable, and providing anti-allergy bedding, upholstery and carpeting.

The major health effects of biocontaminants within buildings have been summarised by Flannigan & Morey (1996) as follows:

❑ Bacterial, fungal and viral infections. Legionnaire's disease, for example, is a bacterial infection typically associated with moisture emitted from cooling towers. This and other ailments, some of which are listed below, have been associated with sick building syndrome (London Hazards Centre, 1990).
❑ Allergic respiratory disease (e.g. extrinsic allergic alveolitis).

❏ Humidifier fever.
❏ Atopic allergic/contact dermatitis.
❏ Endotoxic/mycotoxic effects.
❏ Microbial volatile organic compounds (e.g. from common house dust
   mites).

Influences on dampness

The incidence of dampness in buildings is determined by a variety of
factors. The two primary influences are environment and construction.
Environmental influences are determined by climate, exposure, orien-
tation, ventilation, compartmentation, and building services. Constructional influences are determined by the techniques and materials
used, the quality and suitability of design, the standard of workmanship,
and the level of maintenance.

Environmental influences

The geographical location of a country will of course condition its
climate. The climate is the primary external environmental influence.
Great Britain, for example, can be classed as a northern European
maritime nation.

A climate in a northern latitude tends to be wet, with low sun, a short
summer containing long daylight hours, and not particularly warm. A
maritime region usually has a climate with a high wind velocity, severe
exposure and chill factor. Because of the proximity to seawater, chloride-
contaminated spray as well as fog, mist, and damp, changeable conditions are prevalent. In addition, ivy and other vegetation on the face of
buildings may give them an attractive 'rural' look, but such plants are
hygroscopic and may keep the walls on which they are fixed persistently
damp. All these external environmental conditions will affect a building's exposure or susceptibility to dampness.

The internal environment of a building (i.e. its cryptoclimate) can also
influence the incidence of dampness. Nowadays, improved comfort
conditions are demanded for most building uses, particularly residential
and institutional. The resulting cryptoclimate effects are higher room
temperatures and higher air moisture contents. These, combined with
draught-proofing measures, help to retain more moisture within new
buildings. As a result, such buildings are not allowed to 'breathe'
properly, unless properly ventilated.

Constructional influences

The form of construction can be an important factor in the incidence of dampness. Old buildings, particularly those built before the end of the nineteenth century, generally rely on allowing moisture that has been absorbed by the fabric to evaporate from the surface. They were usually built of solid, load-bearing masonry construction, which more often than not had no provision for damp-proofing within the fabric.

Few dwellings built today have open-hearth fires. The heating of buildings by this method has decreased for three main reasons:

❑ Changing living habits and working patterns have resulted in more homes being vacant through the day.
❑ The Clean Air Act, with its provisions for the setting-up of smokeless zones within urban areas, mitigates against the use of open fires.
❑ There are now more convenient, cheaper, and environmentally friendly heating methods and fuels available.

Moreover, modern buildings depend more on an impervious outer layer or a system of barriers (such as dpc's and dpm's) to prevent moisture from penetrating the fabric. Consequently, they have to rely on good detailing and sound workmanship.

Certain constructional features are more susceptible to dampness than others:

❑ thin external solid walls, particularly of porous materials such as limestone, lightweight concrete blocks;
❑ flat roofs: very shallow pitches are susceptible to ponding, and cold-deck flat roofs (see Chapter 5) are prone interstitial condensation problems;
❑ ground-supported solid floors: failure in or absence of a proper damp-proof membrane can give rise to dampness;
❑ unventilated areas: bathrooms, kitchens, and roof spaces are prone to condensation problems;
❑ suspended ground floor crawl spaces;
❑ hidden gutters at parapet walls and valley gutters;
❑ clipped or flush eaves of pitched and flat roofs (see Fig. 5.1(b)).

Sources of dampness

Three major moisture source categories have been identified (Christian, 1993):

❑ the outdoors (via air and ground);
❑ indoor activities and conditions,
❑ wet construction materials.

These can introduce moisture into a building in a variety of ways (Fig. 1.2).

Most troublesome cases of dampness in buildings usually emanate from one or a combination of three causes: condensation, rising damp and penetrating damp. These will be examined in Chapters 6, 7 and 8 respectively.

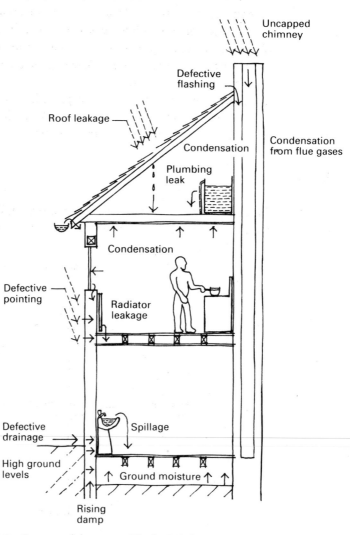

Fig. 1.2   Sources of dampness (Singh, 1994).

There are, however, other sources, which may be less common than those listed above but can still give rise to serious dampness problems. Typical examples of these are:

❑ construction moisture – in new and remodelled buildings – wet floor slabs, screeds, plaster, mortar, and timber, etc.;
❑ leaking services – burst pipes, defective pipe joints;
❑ deliquescent salts – calcium chloride in mortar and concrete;
❑ spillage – cleaning/washing actions; careless filling of liquid containers;
❑ flooding – rise in water table; burst main; overflowing river/stream;
❑ seepage – underground water sources – streams and wells;
❑ sea sand contamination – contaminated aggregates in mortar;
❑ animal contamination – urine contamination on floors and walls;
❑ industrial contamination – leaching of contaminated water from drains and containers;
❑ chimney damp – condensed flue gases from uninsulated chimneys;
❑ floor coverings – deterioration of magnesium oxychloride floors, which have broken down into chlorides.

Moisture sinks

In nature everything tends towards a state of equilibrium. A water source needs to be balanced by a moisture sink. A moisture sink is a medium to

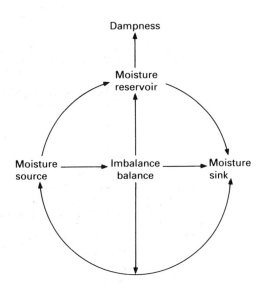

Fig. 1.3 Moisture transfer cycle.

which water can safely dissipate without adversely affecting the building. When this does not take place a moisture reservoir is formed. This cycle is illustrated in Fig. 1.3.

Typical moisture sinks are shown in Fig. 1.4.

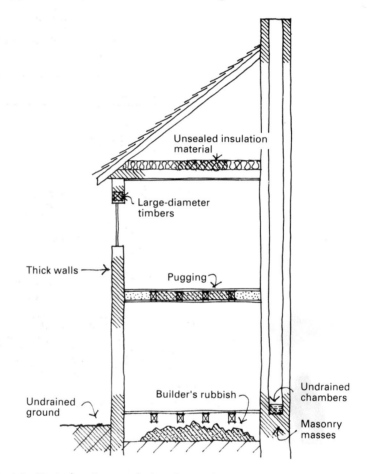

Fig. 1.4   Typical moisture sinks (Singh, 1994).

Moisture reservoirs

Moisture reservoirs are the parts of a building that act as temporary or semi-permanent receptors for moisture. They inhibit or prevent moisture from dissipating to a moisture sink. Typical examples of these reservoirs are hygroscopic materials such as structural timbers, masonry, and insulation. It is the reservoirs in buildings that are susceptible to dampness problems. Examples of moisture reservoirs are illustrated in Fig. 1.5.

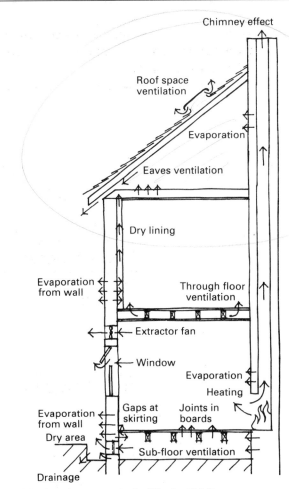

Fig. 1.5   Typical moisture reservoirs (Singh, 1994).

Measurement of moisture

Methodology

A basic requirement of any scientific approach is that the subject or object being studied should in some way be capable of being measured. Moisture is no exception to this rule, if we want to assess and compare its effects on materials and on people. Before it is possible to measure anything, however, it is important that we know what it is that we are trying to measure. With the exception of wood, percentage moisture contents are not meaningful, because they do not tell you if a material is dry or wet (Oxley & Gobert, 1994). There are a variety of means by which this can be done in relation to moisture, and these are considered below.

Moisture in the atmosphere

The moisture status of the atmosphere is usually expressed in terms of humidity. There are three types of humidity measurement: absolute humidity, specific humidity, and relative humidity. The first two are used in meteorology, while the third is more applicable for dampness assessment in buildings. The significance of the various levels of RH can be seen in Table 1.2.

Table 1.2   Some critical RH levels.

RH (%)	Typical effects
100	Saturation percentage
>96	Mould can develop on glass wool
>90	Bacteria can multiply; mould can appear on brick and painted surfaces
>85	Dampness stage: materials may become visibly damp or damp to touch. Timber decay occurs
>76	Mould can develop on leather. Multiplication of mites greatest above this level
>70	Viability of mould increases markedly
65	Maximum optimal comfort level for humans
40–50	Minimum survival level for house dust mites
45	Minimal optimal comfort level for humans. Electrostatic shocks more likely below this level.

The importance of temperature in this regard, of course, should not be overlooked: the higher the temperature, the greater the biological and chemical reactions. Moreover, there is a hygrothermal link with many building defects: for example, melting snow causing damp penetration through the fabric. Hygrothermal performance is defined by Cook & Hinks (1992) as the dynamic interrelationship between the internal and external environments. Many of the dampness problems addressed in this book are hygrothermal in nature.

Moisture in solids

Water activity

The moisture status of non-metallic or non-plastics materials can be described in a number of ways, and is mainly dependent upon the discipline of the measurer involved. Scientists such as microbiologists describe the water availability of a material in terms of water activity (a_w). This is the ratio of the vapour pressure exerted by moisture in the material to the vapour pressure of pure water at the same temperature and pressure. The RH of the atmosphere in equilibrium with material with a particular moisture content (MC) is known as the equilibrium relative humidity (ERH). This is the vapour pressure of the moisture in the material expressed as a percentage of the saturation vapour pressure of pure water. RH is the straight ratio of a_w (Flannigan & Morey, 1996).

Dampness spectrum

As can be seen from the dampness spectrum in Fig. 1.7, a_w can be equated with RH. The spectrum also shows that different materials at the same RH or a_w differ in their moisture content (MC), because of differences in their chemical composition and structure. For instance, at an RH of 80% (a_w = 0.80), the MC of softwood is about 17%, wallpaper 11.3%, cement render 1%, bricks 0.1–0.9%, and Carlite (gypsum) plaster 0.7% (Flannigan, 1992).

In contrast, building engineers and surveyors will tend to use for timber fibre saturation point (maximum sorption content), and for masonry equilibrium moisture content (EMC) or MC to describe the moisture-holding properties of building materials (Flannigan & Morey, 1996). The EMC is that which a material will achieve when it is in balance with the moisture content of the surrounding air.

The basic formula for EMC of a material sample may be written as:

$$\text{EMC (\%)} = \frac{\text{Wet weight} - \text{Dry weight}}{\text{Dry weight}} \times 100$$

For example, if a brick sample weighs 2603 g initially (i.e. when wet) and 2386 g when dry, the difference of 217 g is the mass of the moisture in the material originally, and its total moisture content (TMC) would be

$$\frac{2603 - 2386}{2386} \times 100 = 9\%$$

However, one has to be very careful about using this bare TMC figure. Whether such a material that has a 9% moisture content in wall suffering from, say, rising dampness will depend on the extent of its capillary moisture content (CMC) as opposed to its hygroscopic moisture content (HMC). (See Appendix E.)

Wood moisture equivalent

Another way of describing the level of moisture in building materials is by using wood moisture equivalent (WME). WME is the moisture level in any building material (as if it were in close contact and in moisture equilibrium with wood), expressed as a percentage moisture content of wood (Oxley and Gobert, 1994). It is used in conjunction with certain wood moisture meters, which can also be helpful, initially, to detect dampness in non-timber materials. But, as indicated in Chapter 9 and Appendix M, one should be very careful about using these results on masonry materials at face value because they are often exaggerated. WME can be read on an arbitrary reference scale from 0 to 100, which indicates relative degrees of dampness. The WME readings for various materials listed in Table 1.3 should illustrate the point.

Table 1.3   Typical moisture meter readings on common building materials.

Material	Dry (safe)		Wet (dangerous)	
	EMC (actual moisture content) (%)	WME (meter reading) (%)	EMC (actual moisture content) (%)	WME (meter reading) (%)
Timber	15	15	24	24
Plaster	1	15	3	24
Cement mortar	2	15	6	24
Brick	1	15	5	24

WME readings in excess of 30% should be interpreted with caution. For example, compare two readings taken on a plastered stone wall. One WME reading is 90% and the other, nearby, is 45%. A WME reading of 90% does not necessarily mean that the actual moisture content is double that of the other one taken. What these figures do indicate, though, is that there is a possible dampness problem. They should encourage the surveyor to undertake a more detailed and thorough investigation by (a) undertaking a pinpointing survey of the affected area and (b) taking samples of plaster near the apparent dampness source (Oxley & Gobert, 1994).

Pinpointing survey
This is done by taking readings at regular distances in a grid layout all over the suspect area. These readings should then be noted precisely on a schematic scale diagram (Oxley & Gobert, 1994). The result from such a detailed survey will be a contour of the suspect wall, floor or ceiling, which can help to pinpoint the source of the moisture.

Salts analysis
A small sample of plaster near the affected area can be analysed either by using an on-site kit or by sending the sample to a laboratory to test for the presence of soil salts. If chlorides and/or nitrates are present, then rising damp is the probable cause of the dampness (see Chapter 9 and Appendix D).

Water retention
Water retention is an elaborate subject, and its examination is beyond the scope of this book. However, it is important to appreciate *how* water is retained in materials. In simple terms, with the exception of metals and polymers, most common building materials hold moisture in the following three ways: as chemically combined water, as sorbed water, and as capillary water.

Chemically combined water Water can be held by hydration, or chemically as water of crystallisation in non-organic materials as 'firmly bound water' and, in certain organic substances, as 'water of constitution' (Geary, 1970). It occurs as part of the manufacturing process of compounds such as lime, plaster and concrete. Chemically combined water is permanently held within a material, and is irreversible. It accounts for only a small percentage of the total water held in a material.

Sorbed water A solid material may acquire water by the physical process of sorption. It also includes less firmly 'bound water'. There are three aspects of sorbed water that are important to note: (a) absorption; (b) adsorption; and, (c) desorption. Sorbed water becomes a problem mainly when hygroscopic inorganic salts are activated in high levels of humidity by absorption. Thus a wall not affected by either rising damp or penetrating damp may become wet through these salts.

Capillary moisture This is moisture that moves through the capillaries or pores of a material, and is often known as 'free water'. It is this water that accounts for most of the dampness moisture that affects building materials. It may or may not be held permanently within a material, and its effects are often reversible. (See Chapter 7 and Appendix E.)

Moisture transfer

Moisture can migrate into and within a building in a variety of ways, depending on whether it is in the liquid or vapour state. Moisture transfer is a complex phenomenon involving a variety of processes, such as absorption, evaporation, diffusion, osmosis and capillarity.

The high–low principle propounded by Addleson & Rice (1991) explains the fundamental basis of moisture transfer in buildings (Fig. 1.6).

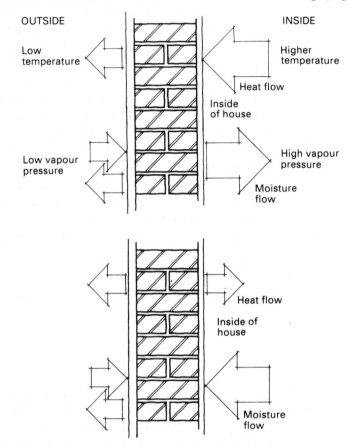

Fig. 1.6   High–low principle of moisture transfer.

What is dampness?

Basis

So far we have ascertained that dampness is unwanted moisture, and we have discussed sources and general effects. To conclude this introductory chapter we need to clarify exactly what is meant by 'dampness'. In other words, we need to have some benchmark from which we can determine whether something is damp or not, and establish a means of measuring it accurately.

Any substance that is not dry could be described as damp. This may serve as a satisfactory definition in general terms, but it cannot be a suitable basis when discussing the presence of moisture in building materials, and indeed in the total building environment. Although many building materials may perform better and prove to be more durable when relatively dry, the comfort of human occupants requires a humidity range that is above that of very dry air (i.e. between 45 and 65% RH) .

The term 'air-dry', however, refers to the condition of a material in an ordinary, inhabited environment with an RH of not more than 70% (Oxley & Gobert, 1994). The biological limit for decay is just below this figure. If a material has a RH slightly higher than 70%, it is becoming damp. Unfortunately, at that level this is not detectable by unaided human senses. It is not until the RH reaches 85% that dampness in some materials becomes visible to the human eye.

The development of dampness problems within the interface between 'air-dry' and 'damp' conditions is not constant for all situations. Biological growths and deterioration may develop just above 70% RH. They are, however, likely to develop more rapidly above 85% RH. Adopting Oxley & Gobert's (1994) approach, therefore, a working definition is:

Dampness can be said to occur when an atmosphere or material is wetter than 85% RH.

Significant dampness

Again, though, one must be careful in using any figure (say, 85%) as a minimum and universally applicable value. That is why the concept of 'significant dampness' is more relevant and meaningful. After all, the severity of any outbreak of dampness is relative. What might be significant or severe in one situation or material may not be in others. For example, a concrete floor slab is considered to be damp if it has an RH above 75% (see Appendix L). Timber, on the other hand, is not wet to the touch and will not be subject to decay until it reaches 85% RH (Oxley & Gobert, 1994).

The significance of moisture in building materials can be expressed in various ways as illustrated in the dampness spectrum diagram in Fig. 1.7.

This book concentrates on dampness in buildings in a relatively wet, temperate climate such as that in the UK. It is hoped, however, that the principles it contains will be of use to those dealing with dampness in other countries.

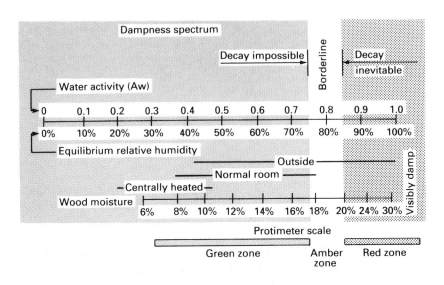

Fig. 1.7   The relations between several different ways of expressing the significance of water in material are illustrated in this diagram, which can be called a 'dampness spectrum'. It is drawn on a regular scale of relative humidity (RH) from zero to 100% and exactly corresponding to this is a scale of 'water activity' (a_w) from zero to unity, which is a very useful measure of wetness and dryness much used in industry. Wood moisture content (average soft wood) is shown, and the correspondence between moisture content and relative humidity is easily seen. The 'visibly damp' indication is very approximate. (After Oxley & Gobert, 1994).

Chapter 2
Dampness and Masonry Products

Movement

This chapter concentrates primarily on stone and clay products. It also, however, considers a phenomenon common to all building materials: movement.

The primary forms of movement affecting a building are:

❑ structural (e.g. overloading);
❑ physical (e.g. vibration, freezing, ground movement);
❑ chemical (e.g. sulphate attack, corrosion);
❑ thermal (e.g. expansion due to increases in temperature);
❑ moisture (e.g. wetting and drying).

For the purposes of this book, of course, it is moisture movement that is being considered here. Readers are referred to the two articles in the Architects Journal on 'Construction risks and remedies: movement' (Anon, 1987a), for a good overview of movement in buildings generally.

Moisture movement is one of the most common causes of cracking, detachment, disfigurement, and subsequent deterioration in building materials. Generally, apart from the special case of frost attack, moisture movement can be divided into two categories:

❑ irreversible movement;
❑ reversible movement.

Many building materials change their size according to whether they are wet or dry. Metals, and other materials such as plastics, are clearly non-absorbent, and will thus not have any moisture movement. Such movement in hygroscopic materials such as timber and wood-based products is generally reversible. Masonry and concrete materials have some reversible movement characteristics but to a lesser degree.

Irreversible moisture movement by its very nature will tend to be more problematic. Its effects are permanent, and often are deleterious.

Chemical reaction is the most common cause of irreversible movement. However, irreversible moisture movement occurs during the initial drying (i.e. shrinkage) of concrete and other cement-based materials.

According to Handisyde (1967):

'Clearly both reversible and irreversible movements of these kinds (i.e. initial drying shrinkage, etc.) must be allowed for, and certain principles stand out:

❏ The magnitude of probable movements must be known if proper provision is to be made to deal with them.
❏ The initial irreversible moisture movement is often much greater than subsequent reversible movement.
❏ The movements of two adjoining materials must be considered together.
❏ Minor movement can sometimes be overcome by means of adequate restraint.
❏ Where appreciable movement is likely to occur, a proper division of the material into small units is essential.'

The moisture and thermal movement characteristics of common building materials are shown in Table 2.1. See also Phillipson (1996) for an overview of the literature on the effects of moisture on porous masonry.

Before looking at the principal masonry materials in more detail, it is helpful to compare the ways in which they deal with moisture in both traditional and modern forms of walling. In Fig. 2.1 a standard solid wall reacts differently from a more conventional cavity wall, but ideally they should both perform adequately in terms of weather resistance.

Natural stone

Types of building stone

Building stones are classified in a similar fashion to the rocks from which they are derived. Because of the great variety of geological formations in the UK (Fig. 2.2), and the diversity of processes that have occurred during geological time, there is much variety in the properties and performance of building stone across the country.

Over the centuries, there has been a natural tendency for buildings to utilise local materials, prior to the availability of the relatively cheap transport of today. Buildings of 'vernacular architecture' are typical examples of the use of such indigenous products. Thus particular parts of

Table 2.1 Moisture and thermal movements of typical building materials (from BRE Digest 228).

Material	Coefficient of linear thermal expansion (per °C × 10^{-6})	Reversible moisture movement (%)	Irreversible moisture movement[a] (%)
Natural stones			
Granite	8–10		
Limestone	3–4	0.01	
Marble	4–6		
Sandstone	7–12	0.07	
Slate	9–11		
Cement-based			
Mortar and fine concrete	10–13	0.02–0.06	0.04–0.10 (–)
Dense-aggregate concrete	12–14	0.02–0.06	0.03–0.08 (–)
Aerated concrete	8	0.02–0.03	0.07–0.09 (–)
Brick/blockwork			
Autoclaved concrete	8	0.02–0.03	0.05–0.09 (–)
Dense aggregate	6–12	0.02–0.04	0.02–0.06 (–)
Calcium silicate	8–14	0.01–0.05	0.01–0.04 (–)
Clay	5–8	0.02	0.02–0.07 (+)
Clay tiling	4–6	No data available	No data available
Metals			
Cast iron	10		
Mild steel	12		
Stainless steel	18 (austenitic)		
Aluminium and alloys	24		
Copper	17		
Lead	30		
Softwood	4–6 with grain 30–70 across grain	0.6–2.6 tangential 0.45–2.0 radial	
Hardwood	4–6 with grain 30–70 across grain	0.8–4.0 tangential 0.5–2.5 radial	

[a] (+), expansion; (–), shrinkage

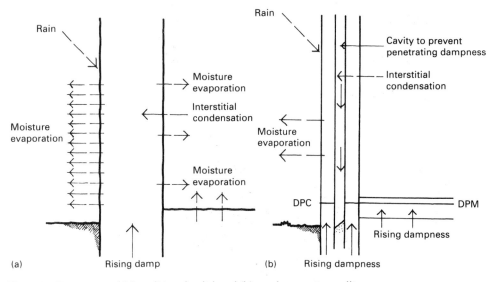

Fig. 2.1   Responses of (a) traditional solid and (b) modern cavity walling.

the country will have developed much experience of the use and performance of their local stone: for example, granite in Aberdeen. Of course, some special building stones of special characteristics and appearance have been selected for prestigious properties in areas far away from the actual site of origin of the stone.

Building stones can be classified into three groups:

❑ igneous – granite;
❑ sedimentary – sandstone and limestone;
❑ metamorphic – marble, slate.

Igneous stones

Igneous stone is derived directly from rock formed from volcanic activity and the solidification of molten material. The commonest example of such stone in the UK is granite, a light-grey-coloured coarsely crystalline acid rock. Basalt, diorite and serpentine are also used. Such stone is mainly employed in the areas where it is quarried, especially in the North of Scotland, the Lake District and Cornwall. Igneous stone is a hard, heavy, dense and strong material with a low degree of porosity. It also has a low water solubility and consequently exhibits excellent weathering properties and durability.

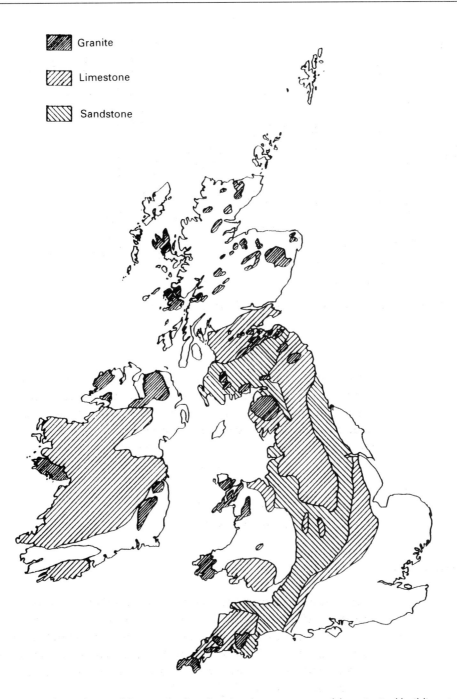

Granite

Limestone

Sandstone

Fig. 2.2 Geological map of the British Isles, showing the occurrence of the principal building stones.

Sedimentary stones

Sources

Sedimentary rocks suitable for building predominate in the North, Midlands and West of the UK, and are found in deposits running in a south-west/north-east direction. Thus a journey west from London, after having passed through the chalk of the Chilterns, would move into the Jurassic limestone areas providing Cotswold stone. Further west Triassic, Devonian and New and Old Red Sandstone areas are found. In other westerly areas, carboniferous limestone, millstone grit sandstone and even older sedimentary rocks are found, juxtaposed among the more recent (geologically speaking) igneous granite (Fig. 2.2).

Sedimentary stones are derived from the products of weathering and erosion of igneous rocks and fragments of organic origin (like coal), which are consolidated by pressure, and the deposition of cementitious matter through the agency of percolating water.

Sandstones

Sandstones are derived from the more resistant constituents of igneous rocks, mainly quartz (pure silica), feldspar and mica (Fig. 2.3). They are part of the arenaceous group of sedimentary stones. The weathering properties of sandstones are, in the main, attributable to the cementitious materials present. The degree of compaction (bedding) is also important, as this will affect the hardness of the stone. Soft-bedded stones, for example, decay quite rapidly (Ross and Butlin, 1989).

The majority of sandstones are frost resistant, and their general weathering characteristics are affected by susceptibility to attack by acidic rainwater containing carbon dioxide or sulphur dioxide. These are the two main acidic atmospheric gases. Calcareous stones are particularly susceptible in this respect. The loss of only a small amount of calcium carbonate matrix (by conversion to calcium sulphate) will loosen a large number of the sand grains in the stone, and rapid deterioration follows. Improperly bedded sandstone blocks will be highly prone to delamination from frost attack and shear stresses (Davey *et al.*, 1995).

Sometimes incompatibility problems occur when sandstones and limestones are juxtaposed to each other in a building to produce colour or variety in a facade or an impression of strength. For example, the products of any acidic rainwater attack on the limestone, principally calcium sulphate, may penetrate the sandstone. Disruption can be caused to the sandstone by the crystallisation of calcium sulphate salts when the percolating water dries out.

In general, sandstones perform well when exposed to weathering. The

(a)

(b)

Fig. 2.3 Diagrammatic representation of the structure of (a) sandstone and (b) oolitic limestone from which porosity could be assessed (× 27 approx.; after BRE Digest 269).

buff/grey sandstones tend to be more durable than some red ones. One of their drawbacks is that they can acquire a coating of soot or dirt in smoke-polluted atmospheres, particularly in urban areas, which is not removed by normal exposure. This soiling usually spoils the general appearance of the building.

Limestones

Limestones have been formed from calcium carbonate deposited from solution in water, either chemically as in oolitic limestones (Portland or Bath stone) or through the action of living organisms that used calcium carbonate for the formation of skeletal structures, as corals and molluscs do today, to form a shelly limestone (Clipsham, Hopton Wood). The original deposits consist of loose particles, that over time became cemented together to form coherent rocks. The chief agent of this cementation action is calcium carbonate deposited from the percolating water.

Oolitic limestones are aggregates of small rounded particles of crystalline calcium carbonate. These have a characteristically concentric structure, consisting of layers of calcium carbonate deposited around a nucleus, which may be a sand grain. The size of the oolite grain varies, and this affects the amount of intergranular cement and the extent of porosity. In some stones the grains are very porous, and in others very dense. Other oolitic limestones have grains that are merely cemented at their points of contact, although intergranular calcite may occur sporadically (Fig. 2.3). In others again, shelly fragments may be found occasionally, or sometimes predominantly, and Barnack limestone is an example of a limestone of this intermediate character.

In addition to calcite, some limestones contain silica deposits. In other limestones magnesium carbonate is found, and these are known as magnesium limestones. The term 'dolomitic limestone' is used to describe limestones that contain a double salt of magnesium and calcium carbonate. Sometimes magnesium limestones contain magnesium sulphate, and this can cause stone decay if it becomes saturated.

Chalk

Chalk is a pure form of limestone. It has been used as a building stone in the south and east of England, where it occurs naturally. It forms a soft white stone, which is rather perishable when exposed to the weather, frost or smoke. A similar stone, Kentish rag, has proved to be more durable to weather exposure.

The lower chalk strata produce a more gritty-textured chalk, which is usually described as clunch. Sand, calcium sulphate and shelly fragments are responsible for the gritty composition. It is tougher, harder, and much more difficult to work than ordinary chalk. It is still quarried at Beer in Devon.

Flint

Flint has been used as a building stone for many centuries in the areas where it is found deposited in Cretaceous chalk downland areas, such as East Anglia and south-east England. Flint is a dark stone of very irregular shape, and consists of nearly pure silica, which develops a white skin on the surface after long exposure. It is a very hard and water-resistant stone, but it can be fractured in any direction. Nowadays flints are mainly crushed for concrete aggregate.

Since some time between the seventeenth and eighteenth centuries, it has become common practice in certain areas to fracture or 'knap' flints to expose the black centre of the stone, and to use these surfaces as the face of the wall to produce a smoother surface. In more prestigious buildings the flints were squared as well, and this produced a stronger structure. An alternative was to use the stones in courses.

Generally, though, flints are of such irregular shapes and sizes that many walls contain more mortar than stone, and because much of this would be lime based, the pace of construction would be very slow. The flint facing might be little more than a veneer, and larger stones, bricks or tiles might be used as bonding units to 'tie' the 'skins' together. Particular difficulties arise at wall angles, at corners or around window and doors, where stone and brick are often used.

The erosion of the lime mortar frequently used in flint walling causes the mortar joints to become furrowed and recessed and more vulnerable to rain penetration. To help overcome this, it was sometimes the practice to insert small flint slivers into the mortar while it was still wet. Known as galleting, this practice gave extra weather protection, and helped to stabilise the larger stones before the mortar set. Although galleting is usually regarded as an ornament, the original purpose was for weather protection.

Chert is a siliceous microcrystalline stone similar to flint, which is found in the West of England. The nodules are paler in colour and larger than those of flint.

Cobbles and pebbles

Cobbles and pebbles found on the seashore or around river estuaries consist of flint or other stones, rounded by the action of water. Walls were constructed with pebbles (less than 80 mm) and cobbles (more than

80 mm), using lime mortar and rubble filling in a similar manner to flint walls.

Metamorphic stones

Slate

The only metamorphic stone that is in use as a general building material is slate. This type of stone is formed as a result of the development of enormous lateral pressures and heat on underground clay deposits. It is a hard fine-grain clay stone with a laminated structure giving parallel planes of cleavage, so that blocks can be easily separated into thin sheets as slates. These have great resistance to water, and this has led to their widespread use as roof coverings, vertical tile hanging and damp-proof courses.

In some parts of the country slate is used as a structural support: for lintels, for example. It has also been used as a walling material, especially in the Lake District and parts of Devon.

The durability of slate varies, even for slate from the same quarry. It depends mainly on the resistance to acid attack, particularly deriving from sulphur dioxide atmosphere pollution, which reacts with slates containing appreciable contents of calcium carbonate. Calcium sulphate is produced, which becomes deposited as crystals that can cause delamination of the slate, especially in covered areas – between overlapping slates, for example – where acid-contaminated rainwater can be held by capillarity. When damp, slate is susceptible to rust marking or algae and moss growth. The former can be difficult to eradicate; the latter may be removed temporarily, but is likely to return in due course.

Marble

Marble is a metamorphic stone that is not used as commonly as slate in buildings. Because of its high cost, marble is usually restricted to use as a superior-quality internal cladding or slabbed floor finish. Marble is virtually impermeable, but discoloration can result in damp conditions, and the backs of slabs must be sealed with shellac to protect them from 'wet' constructions.

Artificial stone

Artificial or reconstituted stone consists of crushed fine aggregate stone mixed with cement and water. It is normally used as a facing to a composite block with a lightweight concrete backing. The reconstituted

stone face is about 50 mm thick, and is attached to the c
with non-ferrous ties.

Although they are not as durable as their natural counterp.
cial stone blocks can provide a relatively cheap way of achieving a..
attractive stone-like face to a building. They can also be used as a decorative cladding on porous brick walls and where a reduction in water
penetration is required.

The mortar used should not be stronger than the background masonry
(i.e. the artificial stone face or its concrete backing), as otherwise thermal, moisture or settlement movement may cause cracking, and penetrating dampness could develop.

Porosity

The porosity of a material is a measure of the total pore space within a
body. However, high porosity does not necessarily indicate a high permeability, as the pores may not be interconnected. The contrast in
appearance between a denser granite or sandstone and an oolitic limestone is obvious when these two stones are examined under the microscope (Fig. 2.3). However, the proportion of coarse and fine pores and
their distribution are also important factors. Thus the stone with a high
proportion of fine pores, sometimes referred to as micropores (up to
0.005 mm in diameter), may be more susceptible to frost damage if
saturated, because the ice crystals have less space to expand into. Coarser
pores with larger diameters may be more accommodating. However,
laboratory experiments on porosity have not always been very successful
in predicting the performance of stone.

Salts

While the porosity of a stone will probably influence frost resistance, it
will also affect susceptibility to damage as a result of the penetration and
crystallisation of soluble salts. This is likely to be a particular problem
with stones with a high proportion of fine pores. In fact, the principal
cause of decay of natural stone is associated with the movement by
moisture of soluble salts and their subsequent deposition when drying
out occurs. These salts may originally be present in the stones themselves, or they may be introduced by other building materials used in the
construction (e.g. mortar sand contaminated by salts), or may even be
formed by the action of atmospheric pollution.

The principal agent of atmospheric pollution is sulphur dioxide. It is an acidic gas, often derived from the burning of coal and other fossil fuels, which naturally contain sulphur impurities. When dissolved in water, sulphurous and sulphuric acid can be formed. This, in turn, can attack limestone and form calcium sulphate. Carbon dioxide is liberated by this reaction and, by forming carbonic acid in the penetrating water, can further attack the limestone. All limestone and calcareous sandstones can be attacked in this way.

The deposition of mineral salts on or near drying surfaces and the subsequent crystallisation of these salts can be a significant factor in stone decay. In buildings where sandstone is used adjacent to limestone, deposition of calcium sulphate can be a major factor. This is another form of incompatibility. Laboratory testing of frost and crystallisation resistance is sometimes a useful guide to the durability of a particular stone as well as of brick.

Bricks

Background

It is believed that the burning of clay to produce bricks was first practised in the Near East about 5000 years ago (Taylor, 1992). However, it was the Romans who first introduced brickmaking into Britain. During the Dark Ages (c.AD 500–1000), manufacture in Britain ceased, and quite a flourishing trade developed in importing bricks from the Continent. In the fourteenth and fifteenth centuries, brickmaking commenced in East Anglia. Initially as expensive as stone, brickwork increased in popularity with the demands of fashion. The shortage of timber and restriction of its use after the Great Fire of London in 1666 further assisted in the expansion of brick manufacture and use.

The development of the railways in the nineteenth century, and the mechanisation of the manufacturing process, resulted in the general availability of bricks throughout the country. They became common in areas where a good local building stone was freely available, so even today Britain can be described as a 'land of brick'.

Manufacture of clay brick

Many different clay materials are suitable for brickmaking. The manufacturing methods vary in detail as a result of the differences in the physical nature of the clays, but all the methods follow a basic sequence of production:

(1) Extract the raw material from the quarry or pit.
(2) Prepare the clay into a consistent plastic paste.
(3) Shape into brick units by moulding, pressing or extrusion.
(4) Dry to reduce moisture content of clay.
(5) Fire to fuse the material into a semi-vitrified mass.

The method used for shaping the brick units dictates the required water content of the prepared clay: moulding methods require a more plastic consistency than do extrusion or pressing. The water is then dried out prior to firing, leaving pores within the material. The ultimate density and crushing strength of the brick are influenced by this factor. However, neither density nor crushing strength are consistently related to the weathering properties of clay bricks (BDA, 1974).

The frost resistance of a clay brick is principally dependent on the mineral nature of the original clay. Temperature and duration of firing are important in relation to the weathering properties of the brick. Firing temperatures vary from 1000 to 3000 °C, and during firing the material is melted and fused into a homogeneous mass. Underfiring leads to poor fusion, resulting in weak bricks and the retention of a high proportion of any soluble salts originally present in the clay. Such bricks should only be used for internal work that is subjected to low stresses; they are not suitable for use in damp or wet conditions because of their salt content and overall weakness. Traditionally, such bricks were used in internal walls until the production of cheap concrete blocks provided an economic alternative.

Underfired bricks were sometimes inadvertently used externally, such as in rendered boundary walls or as the external leaf of rendered cavity walls. In these situations the few underfired bricks in the construction would suffer easily from frost action or salt crystallisation attack. The common symptom of such defects was spalling of the rendering and substrate at isolated bricks.

Flettons are the most common mass-produced bricks, because the clay from which they are made contains material that burns during the firing process. This characteristic therefore saves on the fuel required for brickmaking.

Varieties of clay bricks

Several hundred clay brick products are available in the UK. BS 3921: 1995 *Specification for clay bricks* defines standard terminology and

measurable characteristics. Common facing bricks are, however, not
defined, as concepts of quality depend on appearance only. Common
bricks are for general-purpose construction, with no requirement for
good appearance. Facing bricks are required to be of good appearance,
and are selected to be consistent in the colour and texture of their
exposed faces.

BS 3921 defines crushing strength in units of N/mm^2, and it is used to
predict the loadbearing capacity of brickwork. Water absorption is
measured by stating the percentage weight increase of a water-saturated
brick. Percentage water absorption figures vary between 1.5% and about
30% for very dense and porous bricks respectively. Frost-resistant bricks
are found throughout the range, and water absorption is not always
correlated to frost resistance.

Compressive strength and water absorption are significant in relation
to 'engineering bricks'. In BS 3921, minimum strengths and maximum
water absorbency are specified for two categories of engineering brick
(Table 2.2). Frequently, engineering bricks are frost resistant, but this is
separately defined. Water absorption is also used to define two categories of dpc bricks, which are intended to be used as two courses in
walls with $1:0-\frac{1}{4}:3$ cement:lime:sand mortar.

Table 2.2   Strength properties and water absorption properties of bricks as
specified in BS 3921: 1995.

Type of brick	Water absorption (% m/m)	Minimum average compression strength (N/mm^2)
Engineering class A	4.5	>70
B	7.0	>60
Dpc bricks type A		>5
B		>5
Others	No limits	>5

Durability is designated in BS 3921 in relation to frost resistance
and soluble salt content. In each case durability relates to built brickwork (Table 2.3). Any one brick can have a combination of soluble
salt content and frost resistance. Thus the appropriate type should be
selected according to its location in the building, the local environmental conditions, and its vulnerability to saturation (BS 5268) (see
also Table 2.1).

Table 2.3 Frost-resistant bricks to BS 3921: 1995.

Category	Characteristic	
F	Frost resistant	Resistance to damage even when used in positions liable to saturation and freezing
M	Moderately frost resistant	Suitable for walling protected from saturation
O	Not frost resistant	Internal use only

Tests for frost resistance

The relationship between frost resistance and the saturation coefficient is currently being studied by the BRE to determine whether performance can be predicted. Saturation coefficients are obtained by calculating the relationship between the amount of water absorbed after 24 hours' immersion in water and the amount absorbed after boiling in water for 5 hours, or soaking in a vacuum. Large pores are filled more easily than small ones, the latter allowing expansion of water in the brick upon freezing. It is suggested that a brick with a low saturation coefficient might have more resistance to frost damage, but the practical correlation of these studies has not yet been achieved.

Laboratory tests have been developed by the BRE in an attempt to simulate the weathering of brickwork in buildings. Freeze/thawing cycles of saturated brickwork have been investigated, but correlation with actual performance is not yet satisfactory.

Recent research work has suggested that the shape of the pores in clay bricks may be significant in frost resistance. Bricks with a greater proportion of 'loop', 'blind alley' and 'pocket'-shaped pores have a better frost resistance because they never become completely saturated with water expansion upon freezing. Other shapes, such as 'channel' and 'sealed', become filled more easily, and freezing could damage the brick structure.

Soluble salts in clay brickwork

Under protracted conditions of saturation, soluble sulphates present in some bricks can cause breakdown of cement mortars by causing an expansion reaction, which physically disrupts the mortar. Stronger mortar mixes are more resistant, as they are less permeable and have greater strength. Alternatively, sulphate-resisting cement can be used (see Chapter 3).

Selection of bricks with low soluble salt content (L in BS 3921) should also reduce the chance of such sulphur attack developing. Efflorescence on bricks can occur after wetting, during the drying-out process. Such salts may originate either in the bricks or in the mortar (e.g. by using beach sand contaminated by chlorides). Although it can be unsightly, efflorescence is normally harmless to properly fired bricks and good quality mortars, and usually weathers off in time.

Terracotta

Bricks made from a fine clay that is mixed with sand and carefully fired at high temperatures are known as terracotta. They may or may not be glazed. Sometimes the name 'faience' is used to describe glazed ware. Terracotta is very durable and weather resistant, and in most cases any deterioration can be associated with salt crystallisation. Any cutting or splitting of the bricks is also detrimental.

Calcium silicate (sandlime or flintlime) bricks

Calcium silicate bricks may be less familiar than their clay counterparts, but they are widely used in building, and constitute products of good frost and weathering resistance (BRE Digest 157).

They are manufactured from sand or crushed flint, reacted with hydrated lime and water to form calcium hydrosilicate, which envelopes the sand particles and bonds them together. The constituents are mixed, compressed into a brick shape, and then autoclaved under steam pressure for several hours. Properties of compressive strength and porosity vary according to processing procedures. Compressive strengths range from 20 to 50 N/mm^2, and all such bricks are regarded as frost-resistant products. They do not contain soluble salts, so do not contribute to efflorescence or sulphate attack.

However, calcium silicate bricks have a higher coefficient of linear thermal expansion (see Table 2.1) than most clay bricks. Brickwork constructed using calcium silicate bricks therefore must contain vertical expansion joints at 9 m centres (4.5 m in parapets), rather than the 12 m (6 m in parapets) centres required for clay brickwork.

Clay tiling

Clay roofing tiles are manufactured in many parts of the country where suitable clays are available. The process involves the extrusion of clay through dies. This clay is then fired in a down-draught kiln at temperatures over 1000 °C. Tiles made from uturia marl clay have an excellent reputation for durability, and this is probably due to their low water absorbency – not more than 5% on the BS 402 water absorption test, compared with other clays that may have up to 10% water absorbency in this test. Pantiles are kilned at lower temperatures, and may be even more water absorbent (14%).

Greater water absorbency may lead to frost damage, especially on roofs with higher standards of insulation, which in effect lowers the temperature of the roof tiles during cold weather periods. The commonest symptom of frost damage is delamination of the tile. Roof pitches above 35° are recommended for clay-tiled roofs, to reduce the extent of water penetration.

Quarry tiles and other clay floor tiles are very durable and have good resistance to chemicals. However, like roof tiling, they can suffer from frost damage if exposed to cold damp conditions.

Unfortunately, at the time of writing, data were not available on either reversible or irreversible moisture movement of clay roof tiling.

Mud and earth walls

In certain parts of the country, buildings will be found with walls of clay. Various systems of construction were employed, but the best known are cob and whichert. Cob is the name used in Devon and Dorset, whereas whichert is used in Buckinghamshire. The processing of clay and the construction of the walls themselves was laborious and time-consuming. A marl was selected, and sometimes chalk was added to assist setting. The mixture was watered and trodden by horses, and straw was added. The mixture was then watered and further trodden in. A foundation wall of stone, pebbles, flint or brick about 0.4–1.0 m thick was constructed to a height of 0.3–0.5 m above ground. On top of this stone wall, layers of trodden and processed clay were added. Each layer would be about 0.5 m high. When it was dry, another course was laid. When the required height was reached, the wall would be protected by overhanging eaves and coated with lime wash or plaster, or even hung with tiles or slates (McCann, 1983).

An alternative system was the clay lump or clay bat. Clay and straw was formed into rectangular blocks and dried naturally in wooden moulds $0.5 \times 0.3 \times 0.15$ m, usually for two or three weeks before laying. This system is mainly found in East Anglia. Yet another variation is the *pise de terre* method, in which semi-dry earth was rammed between boards in situ on the wall using a specially designed frame that could resist ramming, yet be demountable.

An essential requirement for these types of walls is that they are kept relatively dry (Trotman, 1995). The design of overhanging roofs, rendered faces and high stone wall bases was presumably intended to minimise the effects of rising and penetrating damp. If clay becomes damp, the binding straw will decay and the wall will become weakened and disintegrate. Alec Clifton-Taylor quotes an old Devonshire saying that 'all cob wants is a good hat and a good pair of shoes': that is, a solid plinth and some kind of roof (Clifton-Taylor, 1962). Many mud buildings are still in use today after many hundreds of years. All such buildings are well insulated, and are described by the occupants as being warm in winter and cool in summer: the ideal internal environmental conditions of a building.

The traditional external render for such walls would be a clay- or lime-based render, coated regularly with a coloured lime wash. Such a render would be porous, and would allow the wall to 'breath' and lose any excess water by evaporation during dry weather periods. The use of dense cementitious renders and impervious paints would be undesirable for such construction.

The cellulose base of 'size', a typical matrix in whitewash, will make walls coated with this material susceptible to organic growths.

For a more detailed account of the repairs required for cob and other traditional buildings, see the five-volume series on conservation by Ashurst & Ashurst (1988).

Masonry problems and their therapies

Conditions for soiling and deterioration

Soiling

As we have seen, moisture plays a critical role in the soiling and deterioration of masonry. For example, as regards soiling, the availability of water is probably the most critical factor in allowing the colonisation on a stone surface, and the amount of water determines the species of organism that occurs (Andrew *et al.*, 1994). The overall effect is to darken or stain the appearance of the masonry.

The other conditions required for soiling are:

❑ light – for photosynthetic organisms such as algae;
❑ temperature – neither too hot nor too cold;
❑ pH – highly alkaline substrates deter mould growth;
❑ nutrition – airborne pollen alighting on wall surfaces; organic constituents of coatings (e.g. cellulose in 'distemper').

Building soiling can be categorised into two groups: non-biological soiling (e.g. atmospheric constituents and pollution, weathering, graffiti, soot, oxides), and biological soiling (e.g. algae, fungi, bacteria, lichens and moss). Not only does soiling spoil the appearance of a building, it can also encourage or accelerate the development of failure mechanisms (see below).

Deterioration
Moisture is also a key factor in many deterioration mechanisms from frost action to acid attack. The conditions required other than water are:

❑ exposure;
❑ atmospheric pollution;
❑ selection and detailing of stone;
❑ workmanship;
❑ porosity.

These deterioration mechanisms can lead to a number of, mainly moisture-related, problems or failure mechanisms.

Failure mechanisms

Stone- and brick-faced buildings can suffer from a variety of defects, some of which have already been mentioned. The main symptoms of such problems are:

❑ delamination and spalling – detachment of lamina on incorrectly bedded stones or poorly fired bricks;
❑ soiling and staining – grime, soot and atmospheric pollution;
❑ cracking – moisture etc. movement;
❑ erosion – impinging wind-driven rain; abrasion;
❑ friability – crystallisation of soluble salts; acids produced by pollutants and bacteria, dissolving the calcite.

Therapies

Cleaning of masonry

The frequency of cleaning required depends upon the degree of atmospheric pollution, the exposure and the type of stone. Thus limestone that is freely washed by rain may be 'self-cleansing'; in protected positions limestones and sandstones may require cleaning at 5–10 year intervals and brickwork at 10–20 year intervals, while in polluted atmospheres polished marble may have to be cleaned every month (BRE Digest 280).

The appropriate method of cleaning masonry will depend on a variety of factors, such as the type and quality of masonry material, the condition of the facade, its exposure, the aim and purpose of the cleaning programme, and the attitude of the building's owner towards cleaning the facade. In addition, the policy of the local planning authority and historic buildings agency should not be ignored here.

The cleaning and treatment of masonry soiled or damaged requires a careful and sensitive approach. This is particularly so in the case of historic buildings. In the 1980s and early 1990s a number of ill-thought or badly executed cleaning programmes to some historic buildings have had deleterious effects on the stonework (Andrew et al., 1994).

Masonry cleaning methods

Since about the early 1970s it has become fashionable in some British towns to carry out vigorous cleaning programmes to many stone and brick-faced buildings. The reasons for this intervention are usually aesthetic or conservation influences, or a combination of both. Such treatments, however, have not always proved successful. In some cases stone-cleaning schemes have made matters worse, by bleaching the stone, causing efflorescence and chemical staining, removing details, and accelerating the stone's rate of deterioration.

Deposits on limestones can usually be softened by a fine spray of clean water, and sometimes marbles can be cleaned in this way, but rarely sandstones or granites (BRE Digest 280). Caustic soda and soda ash are very damaging, and must never be used on any stone or brick.

Brickwork can usually be cleaned by a fine water spray or by chemical cleaning using a dilute (5%) hydrofluoric acid solution similar to that used for stone cleaning by chemical means.

Methods of stone cleaning can be divided into the following two main categories:

❑ traditional stone-cleaning methods:
- physical: dry – grit/sand blasting; abrasive wheel;
                wet – steam; high/low water washing;

- chemical: dilute acid/alkali solutions; poultice (Andrews *et al.*, 1994).
❏ modern stone-cleaning methods:
  - dry ice (cryogenesis or 'Drice') method (Thomson, 1994), involving firing grains of dry ice onto the face of the stone;
  - lasers to vaporise pollution from the stone's surface;
  - polystyrene pellets fired onto the face of the stone to remove dirt and grime.

The main benefits of these modern methods are that they use much less quantities of water, they keep abrasion to a minimum, and they are more environmentally friendly, because they are less messy and do not generate as much waste. However, such methods are relatively expensive and, like most blast-cleaning techniques, may not be suitable for softer stones such as limestone or some sandstones. A test patch is always recommended with any cleaning technique (Thomson, 1994).

Repairing defective masonry
Stone or brick that is damaged by chemical or physical attack can be repaired in two main ways:

❏ Plastic repairs – patching defective areas of brick and stone with a suitable cement:sand grout to match the substrate. This method is only suitable for small areas (Fig. 2.4). It should not be adopted when dealing with block-size or large-scale repairs, as it is not a very durable solution.
❏ Indentation – this involves cutting out the defective unit, or at least the first 75 mm from its face. In the case of a defective brick, it should be replaced in its entirety. Stonework, on the other hand, can usually be repaired effectively using this method (Fig. 2.5).

At the time of writing nobody has developed a totally effective technique for consolidating friable masonry (Richardson, 1991). It will probably be many years before a widely accepted chemical consolidant will be available. Also, given the wide range of masonry materials with their different characteristics, it is unlikely that a single consolidant will be universally applicable.

In any event, a number of guiding principles should apply in sensitive or important repair programmes:

❏ Repair like with like: ensure compatibility.
❏ Use tried and tested repair methods rather than unproven innovative techniques.
❏ Minimise intervention: if in doubt, postpone any aggressive masonry

cleaning programme until the most appropriate therapy is determined.
- ❏ Reversibility: any treatment should be reversible.
- ❏ Use only experienced and skilled operatives.
- ❏ Adopt a planned preventive maintenance system for the building.

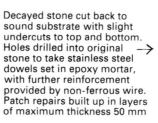

Decayed stone cut back to sound substrate with slight undercuts to top and bottom. Holes drilled into original stone to take stainless steel dowels set in epoxy mortar, with further reinforcement provided by non-ferrous wire. Patch repairs built up in layers of maximum thickness 50 mm

Fig. 2.4   Plastic repair method.

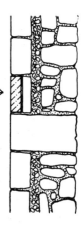

Decayed stone cut out and new stone indent 75–100 mm. thick inserted into wall, bedded and pointed in lime mortar to match existing mortar

Fig. 2.5   Indentation repair method.

Chapter 3
Dampness and Cementitious Products, Bitumens and Metals

Cement

Background

The term 'cement' is loosely used to describe a substance that unites other materials to form a solid whole. In the context of this chapter, however, it will be restricted to substances that can be described as calcareous cements (i.e. based on calcium compounds).

Joseph Aspdin, a Leeds builder, in 1824 patented what he called 'Portland cement' because it resembles Portland stone in appearance (Neville & Brooks, 1987). It was manufactured by heating clay and limestone to drive off carbon dioxide from the limestone. Later modifications to the process included greater heating to form a clinker. Today, most cement is made by a wet process in which clay and limestone are ground to a slurry and then fired. The dried clinker from this process is then ground to a powder in a ball mill. A typical cement mixture is given in Appendix A. The final set cement forms a calcium–iron–aluminium complex.

A wide range of building materials are cement based, and include mortars and renders, mass concrete, precast concrete products, and reinforced and prestressed concrete. An equally wide range of aggregates are employed in the manufacture of these different products. Many additives and admixtures are used to produce a variety of special types of cement and concrete products for specific use.

Most cements are hydraulic (i.e. they set by chemical reaction with water). The basic raw materials for ordinary Portland cement (BS 12) are clay or shale, and calcium carbonate in the form of chalk or limestone (see Appendix A). Most clays are made up of silica, alumina and iron oxide. These constituents are finely ground, mixed and heated to about 300 °C in a rotary kiln, and a clinker is formed on subsequent cooling. Gypsum (calcium sulphate) is added, and acts as a retarder, and the whole mixture is ground to a powder (Portland cement). Details of the chemical changes that occur are described in Appendix A.

Hydration of Portland cements

When cement powder is mixed with water it undergoes a process called hydration, which leads to setting and hardening. Depending on the aggregate, the familiar range of cement products are produced (e.g. mortar, render, concrete, concrete blocks).

Immediately on mixing with water, the cement grains become dispersed in the mixing water. Their distribution will be determined in part by the amount of water added (the water:cement ratio). Some cement components react very quickly with the water, and stiffen the cement immediately the water is added. However, gypsum powder retards this rapid hardening until it is all used up, which usually takes about 24 hours. Hydration of the cement grains now commences, and electron microscope observations indicate that rods and needles of ettringite and calcium silicate form in the space between the cement grains. After 24 hours, a continuous gel is formed and setting is occurring, but no real strength has developed (see Fig. A.1 in Appendix A).

After seven days, considerable strength has developed, although not all the hydration reactions are complete. Many more hydration links have formed, although not all the reactions are complete, and some capillary pores remain unfilled. As ageing takes place, the total pore volume declines as the larger pores are reduced in number.

The amount of water used to hydrate the cement will profoundly affect the final strength of the cement. More water has to be added to make the cement workable than is needed for chemical setting. The remaining water evaporates during the 'curing' process and leaves capillary pores, which may reduce the strength and increase permeability to water (Fig. 3.1).

Aggregates

It is the usual practice for cement products to incorporate an aggregate, which, because of the lower cost, will add bulk to the mix and produce an economically attractive material (BS 882: 1992). The objective is to use as much of this cheaper 'bulk' material as possible while at the same time making a product of the required properties, the cement filling the interstices between the aggregate particles. Thus several sizes of aggregate particles will be used in some products: fine aggregates only in mortar; coarse particles only in 'no-fines'; and fine and coarse aggregates

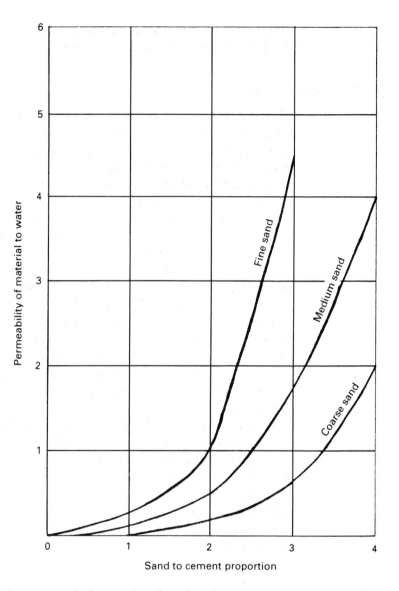

Fig. 3.1   Graph showing the effect of sand particle size on cement rendering permeability (after Ragsdale & Raynham, 1972).

in ordinary concrete. They are usually inert, and do not hydrate significantly.

Concrete

Concrete is the term use to describe an aqueous mix of cement with sand (the fine aggregate) and mineral matter (the coarse aggregate). Although it was used as a building material by the Romans, it was not until the nineteenth century that concrete became a common construction material in the UK.

Sand is an essential ingredient in ordinary mass and reinforced concrete, as it reacts with the calcium hydroxide liberated during the setting of the concrete. In general, siliceous particles with an average diameter less than 5 mm are called sand, while larger particles are referred to as gravel. Crushed rock, granite, basalt, sandstone and limestone may also be added. The quality and grading of the aggregate are important, and the presence of organic impurities can affect both the hardening and the weather resistance of the concrete (Lee, 1980).

The grading of the sand and the size and distribution of its particles are important. Sand contamination by impurities and particularly sands with a high proportion of fine particles often produce concrete products of lower durability (Figs 3.1, 3.2).

In steel-reinforced concrete, it is especially important that moisture and air penetration are kept to a minimum. Concrete of poor quality or inadequate thickness can allow corrosion of the steel reinforcement to occur, and this causes spalling and cracking of the concrete cover. Ordinarily as much as 40 mm cover of reinforcement is now considered necessary (75 mm in substructure or very exposed elements).

Lightweight aggregates, such as blast-furnace slag and fly ash from coal-fired power stations, are also used to lower the costs and modify the properties of some concretes. A range of densities from 400 to 4000 kg/m^2 is possible, and a wide range of strength properties are produced by incorporating various types of aggregate.

Durability of concrete and cement

Deterioration mechanisms

Concrete was once considered to be a relatively maintenance-free material. The many problems of cracking and deterioration of concrete-framed and panelled buildings and bridges have put paid to that impression. Properly designed, installed and used, however, concrete can be a very durable material.

A durable concrete or cement product will only be one that can withstand the effects of destructive processes, which can act externally or internally. The external threats develop from the penetration of aggressive agents. Chemicals may affect the concrete itself or cause deterioration or chemical changes in the metal reinforcement. Problems may also be induced internally by chloride attack of the reinforcement

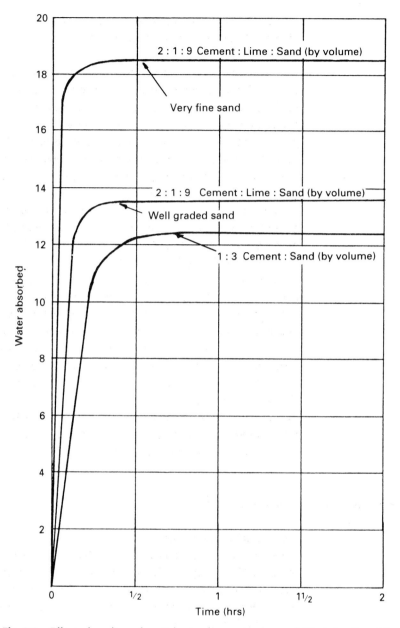

Fig. 3.2 Effect of grading of particle size on mortar permeability (after Ragsdale & Raynham, 1972).

from chlorides used in the mix, or by deterioration of the coarse aggregates by alkali–aggregate reaction.

Clearly, the effects of external agents can be reduced or avoided if the concrete is impermeable or has a low porosity. This is achieved if the water cement ratio is low (i.e. not more than 0.6), and curing should achieve the maximum level of hydration. Full compaction and good workability are also beneficial in reducing porosity. The approximate level of cement content will ensure greater strength, which can better resist the disruptive forces of deterioration. Correct mix design, good concrete practice, and care in batching, mixing, placing and curing, are all important.

It is probably sulphates that represent the major cause of chemical attack of concrete and cement products. They are naturally present in many soils, and also form as a result of atmospheric sulphur dioxide pollution. Condensation of sulphurous flue gases in chimneys and sulphates from gypsum plasters, which may become wet, are also sometimes responsible for causing sulphate attack. Sulphate attack causes cracking and spalling of the concrete, and in severe cases a patch repair is not possible. The best precautions to follow to prevent sulphate attack are to reduce the permeability of the concrete, and to use sulphate-resisting cement.

In reinforced concrete, the steel may rust if the concrete cover is inadequate, resulting in loss of effective cross-sectional area of the steel and increasing the stresses on the remaining unaffected reinforcement. Rusting also results in an increase in volume of the corroded steel's perimeter, and this internal expansion leads to disruption, spalling and accelerated corrosion.

If concrete is kept dry, the combination of water and oxygen that is essential for rusting cannot occur. The high alkalinity of the cement hydrates provides a chemical protection by the formation of a thin oxide film on the surface of the steel. Any fluid that can penetrate and reduce this alkalinity (i.e. increase the concrete's acidity) reduces corrosion resistance as well. Atmospheric carbon dioxide may react with calcium hydroxide in the cement to form calcium carbonate. This 'carbonation' process is more rapid in poor-quality concrete, or if sulphur dioxide is present as an atmospheric pollutant. The process causes shrinkage, increasing water penetration, and reduces the alkalinity of the concrete (BRE Digest 405). This increased acidity of the concrete leads to corrosion and rusting of the steel reinforcement, causing swelling and cracking of adjacent concrete.

Chlorides derived from de-icing treatment on roads, or from the use of calcium chloride concrete-setting accelerators (sometimes used in winter working conditions), can also act as agents of deterioration.

Internal disruption of concrete can occur if temperature reduction causes expansion of any free water as it freezes. If there are no air voids, an expansion occurs and the structure may be disrupted. Subsequent thawing allows additional water penetration and further deterioration when freezing occurs again. This may appear as 'pop-outs' at the surface, pock-marked concrete surfaces, or patterned cracking or scaling of the surface. The expansion of the internal capillary water on freezing may be dispersed by including air voids in the concrete.

The voids in cement products are often divided into three main types: mini pores, which are present in the cement gel, and are too small for capillary action; capillary pores, which soak up water; and voids, which may be several millimetres in size. Voids are good for frost resistance. Capillary pores are caused by high water:cement ratios, and cause excessive porosity. The best precautions require the rejection of highly water-absorbent aggregates, a low water:cement ratio (<0.6), good curing, and the use of air-entraining agent admixture, which introduces air bubbles up to 0.1 mm in diameter.

Sulphate-resisting cement

Sodium sulphate attacks the tricalcium aluminiun hydrate, forming the calcium sulphoaluminate ettringite (see Appendix A), which is accommodated by a considerable increase in volume (over 200%). It also reacts with calcium hydroxide to form gypsum, and this reaction is accompanied by an increase in volume. Sulphate mainly emanates from the soil, but may also come from contaminants in the brick or mortar. The severity of attack depends on the type of sulphate involved and the strength of the sulphate solution, but loss of strength can be very rapid. Again, moisture is the catalyst for this type of chemical reaction.

The best precautions are to use sulphate-resisting cement, which is manufactured to contain a low proportion of tricalcium aluminate (C_3A), limited to 3.5%. By keeping the cement product dry, and reducing permeability, the opportunities for water penetration and hence sulphate attack are reduced.

Alkali–aggregate reaction

Alkali–aggregate reaction (AAR), alkali–silica reaction (ASR) or alkali–carbonate reaction (ACR) (Son & Yuen, 1993), sometimes described as 'concrete cancer', occurs when a series of reactions take place between certain aggregates containing permeable and highly siliceous rocks (e.g. opal) and silicates originating in Portland cement. This reaction forms a gel on the surface of the aggregates, which can absorb water, causing volume expansion (BRE Digest 330). The symptoms of this decay mechanism are spalling and extensive crocodile-like cracking.

High alumina cement

High alumina cement (HAC) was a popular constituent of precast and prestressed concrete in Britain until the 1970s. It has rapid hardening properties and excellent resistance to high temperatures: hence its use in refractory concrete. However, after a series of serious structural failures in buildings with HAC concrete in the early 1970s, it was withdrawn from general construction use (BRE Digest 392). In July 1974 the Department of the Environment indicated that all multi-storey buildings containing HAC must considered as suspect (Hollis, 1991).

HAC, when used in concrete, undergoes a chemical change known as conversion. This is accompanied by a loss of strength and reduction of resistance to chemical attack (Hollis, 1991). This process was aggravated in buildings with high levels of humidity (e.g. swimming pools). As a result, highly converted HAC concrete in such conditions was very susceptible to acid, alkali and sulphate attack. HAC is also vulnerable to chemical attack from plasters and wood-wool slabs. Both white and black markings on the face of HAC concrete may be indicative of such attack.

Precautions

The common factor in all these 'non-structural' causes of concrete decay is moisture. Without moisture there would be little corrosion of steel, no attack by chlorides, sulphates or frost, and little or no AAR or HAC problems. However, given that concrete will be exposed to some moisture during its service life at some time, certain precautions can be taken to enhance its durability:

❏ Use 'pozzolanic' or other micro-silica cement (with grains 100 times smaller than ordinary Portland cement) for increased density of concrete.
❏ Keep the water : cement ratio at about 0.5 or lower.
❏ All steel reinforcement at exposed concrete faces should have a minimum 50 mm cover (75 mm in aggressive conditions).
❏ Protect the finished concrete with a two-coat liquid PVC protective paint system (e.g. Decadex).

Additives and admixtures

Admixtures are materials that are added to part of the cement or other binder during its manufacture. Examples of additives would be cement process grinding aids such as polyethylene glycol. The term is also used to describe those substances that are part of a pre-mix binder/cement during manufacture, such as microsilica or other integral cement particles.

Admixtures, on the other hand, are special materials added in small quantities to cement or concrete mixes on site, which alter the properties or strength of the product. Examples of admixtures are air-entraining agents and special aggregates (Rixom, 1977).

Plasticisers

A stronger and more durable concrete or cement product can be produced by reducing the water content of the mix. Water reducers or plasticisers such as lignosulphonates or hydrocarboxylic acids are used. These products are absorbed onto the surface of the cement grains, which become mutually repellent and therefore become more dispersed. These admixtures have three main functions:

❏ to achieve a higher strength by decreasing the water : cement ratio at the same workability as the admixture-free mix;
❏ to achieve the same workability by decreasing the cement content so as to reduce the heat of hydration in mass concrete;
❏ to increase the workability so as to ease placing in inaccessible areas (Neville & Brooks, 1987).

Air entrainers

Air entrainers are included in the mix (Table 3.1) to produce small air bubbles up to 0.1 mm in diameter. Entrained air must be well and evenly

Table 3.1 Typical air-entraining agents.

Air-entraining agent	% in cement
Alkali salts of wood resins	0.025–0.1
Sodium sulphonate or sulphonated napthenates (detergents)	0.25–0.1
Calcium lignosulphonate	0.25–1.0
Sodium salts of napthenic acid	0.025–0.1
Calcium salts of proteins	0.25–0.5
Alkali or triethanolamine salts of fatty acids	0.025–0.1
Triethanolamine salts of sulphonated aromatic hydrocarbons	0.025–0.1

dispersed. The bubbles must be small for effective frost resistance and to minimise loss of strength. Unintentionally entrapped air may form larger and more irregular voids. This could arise as a result of using irregularly shaped aggregates or inadequate vibration.

The reduction in strength of air entrainment would be about 25%, but the improvement in workability permits the use of a lower water content in the mixing, and this reduces strength loss to about 8% (Neville and Brooks, 1987).

Air entrainers (Table 3.1) can be interground with the cement (i.e. an additive) or added to the mix (i.e. an admixture). They act either by reducing water surface tension in the mix or by producing very small bubbles, like foam on soapy water. This entrainment increases resistance to both frost and sulphate attack. The aim is to entrain 5–6% by volume of air with 25 mm aggregate concrete and 1% in ordinary dense concrete. Mortar entrainers produce much more air, perhaps twice the level used in concrete, to produce increased plasticity. Increased workability is likely to produce greater uniformity of the mix, and permeability is reduced as a result of the lower water : cement ratio. Frost resistance is probably better in air-entrained concrete because the distance between the air bubbles is so small that pressures induced by the growth of ice crystals are reduced.

Integral and surface waterproofers

Many chemicals are added to cement mixes to act as integral water-proofers (Table 3.2). Stearates and silicates are commonly used. These can affect capillary forces and are useful in damp-proofing, but they cannot operate against hydrostatic water pressure.

Surface waterproofers can be formulated for surface waterproofing of in-situ concrete. Bitumen coatings and emulsions, chlorinated rubbers, PVC coatings, synthetic resin lacquers, waxes, silicones and latex coatings are used. Drying oils (Fig. 3.3) can form calcium soaps with free calcium hydroxide in the concrete, and these act as effective water stops. Sodium silicate is also used in this way, and this hydrolyses, releasing silicone dioxide, which reacts with free lime (calcium oxide) to form crystalline structures that can block pores. Magnesium and zinc silico-fluorides act in a similar way. Silicon tetrafluoride vapour produces an impermeable coating of calcium silicofluoride on the capillary surfaces.

Another approach is to employ hydrophobic substances, which line the capillaries and make them water repellent. These include silicones, siliconates and alkali soaps (e.g. butyl stearate and oleate). Some of these

Table 3.2 Typical integral waterproofers.

Type	Example
Inert materials: act as workability aids and pore fillers	Hydrated lime, clays, silica siliconates, talc, chalk, barium sulphate
Pore fillers: active materials[a]	Alkali silicates $\rightarrow$ Calcium silicates Silicofluorides Iron filings and ammonium chloride Diatomaceous silica
Inert water repellents	Calcium soaps Waxes and mineral oils Bitumen emulsions Coal tar residues
Active water repellents[a]	Alkali soaps $\rightarrow$ Calcium soaps Free fatty acids Butyl stearates

[a] Active products react with cement as indicated.

Note: Some of these products cannot operate effectively against hydrostatic pressure, and can only be used to reduce water penetration by capillarity.

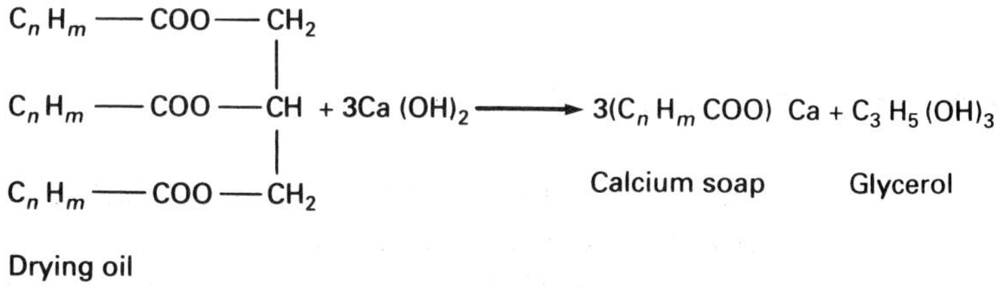

$$C_n H_m - COO - CH_2$$
$$|$$
$$C_n H_m - COO - CH + 3Ca\,(OH)_2 \longrightarrow 3(C_n H_m\,COO)\,Ca + C_3 H_5\,(OH)_3$$
$$|$$
$$C_n H_m - COO - CH_2 \qquad\qquad Calcium\ soap \qquad Glycerol$$

Drying oil

Fig. 3.3 The reaction of drying oils with calcium hydroxide in concrete to form 'water stops'.

chemicals combine pore-blocking and water-repellency properties (Table 3.2).

Concrete blocks

The manufacture of concrete blocks was started by a Mr Ranger, a Brighton builder, who filed a patent for concrete blocks in 1832. He took shingle from the beach, broke flints and stone chippings, and used chalk lime for the cement. Since that time concrete blocks have become widely used, initially as a substitute for stone and to provide a cheaper form of wall masonry. But today a wide range of block types are available, with a

Table 3.3   Properties of concrete blocks and bricks.

Type	Density (kg/m^3)	Compressive strength (N/mm^2)	Aggregate
Dense aggregate blocks[a]	1300–2000	3–10 (20–30 possible)	Limestone Granite Gravel and sand
	800–1200	3–7	Sand Pumice Pelletised PFA
Aerated concrete blocks	480–760	2–3 (7 possible)	Lime Sand PFA Aluminium powder (aeration)
Reconstituted stone	2000	20–25	Limestone
Concrete brick[b]	2150–2300	20–40	

[a] For use in more exposed situations (e.g. copings, parapets and external walls near ground level, blocks of higher compressive strength (7 N/mm^2) and higher density (over 1500 kg/m^3) (BS 5628 Part 3: 1985) or
[b] engineering quality concrete bricks with a higher compressive strength (over 30 N/mm^2) and lower water absorption are recommended.

variety of properties (Table 3.3). Their moisture and thermal properties were compared with those of brick in the previous chapter.

Dense aggregate concrete blocks are suitable for load-bearing masonry work in external environments and below dpc level. They normally consist of limestone, granite or crushed gravel and sand aggregates, and are manufactured in either a solid, hollow or cellular form. They have high strength, high thermal capacity, and a low sound transmission. As can be seen from Table 2.1, they have a moisture movement rate similar to that of clay brick.

Lightweight aggregate blocks can be used for ordinary above-grade load-bearing or non-load-bearing situations. They are usually made from cement, sand, and lightweight aggregate such as pumice or a manufactured aggregate such as pelletised pulverised fuel ash (Lytag). Waste materials such as cinders, coke breeze or furnace slag are also used. Lightweight concrete blocks are easier to handle and lay, and have good thermal and insulation properties. They are easy to cut, chase or shape. Adjusting cement content can affect moisture shrinkage. An open texture is preferred, with cement bonding the points of contact of the aggregate to provide a better mortar key, more uniform weathering and less efflorescence, less capillary action, quicker drying and lower frost susceptibility.

Aerated concrete blocks are manufactured from cement, lime, sand, and pulverised fuel ash (PFA) with aluminium, which is responsible, on mixing, for aeration. As would be expected, they have superior insulation properties, yet some types can be used for load-bearing situations, even below ground level.

Clinker or breeze blocks are manufactured using an aggregate obtained from power station fuel ash. In modern block manufacture, the better grades of PFA or furnace-bottom ash are preferred, as they contain less unburnt organic matter and dust. Blocks with a high proportion of organic material are subject to excessive moisture movement and biological growths. Fly ash collected from flue gases at power stations is cementitious, and acts as a cement substitute in some block-manufacturing processes. Various aggregates are used, including expanded clay, shale, foamed slag and exfoliated vermiculite. The main use for these blocks is for non-load-bearing partition walls above dpc level.

Generally, concrete blocks are not suitable for deep substructures or basement construction where internal tanking is used and high levels of hydrostatic water are anticipated. The same requirements for damp-proof courses apply to blocks as well as to bricks.

Concrete bricks

Concrete bricks are manufactured from a mixture of crushed aggregate and cement, with the colours being varied by adding pigment or mixing the aggregates. They are cured in a steam oven for between 12 and 24 hours, and then allowed to cure for about a month before delivery, by which time most of the initial drying shrinkage will have occurred. Their price compares favourably with that of clay bricks.

Three types of concrete brick are produced: commons; facings (with a variety of colours and finishes, and higher cement content); and engineering (with higher strength and even greater cement content). Engineering concrete bricks are suitable for use as copings, sills, and in earth-retaining situations even where sulphate conditions exist (BS 5628 Part 3: 1985) (Table 3.3).

Because of their relatively higher thermal and moisture movement rates, concrete block and brick walls should have vertical movement joints every 6 m. This compares with 12 m and 7.5 m for fired clay and calcium silicate bricks respectively. All such movement joints should be able to accommodate 10 mm of movement.

Concrete roof tiles

Concrete roof tiles were first manufactured in Bavaria in the 1840s, where a naturally occurring pozzolanic cement and local sand were used. Tiles made at that time were still in use 100 years later.

The introduction of cement and mass production techniques have resulted in concrete tiles having the largest share of the roof tiling market in the UK today. A whole range of different shapes (flat, ribbed, etc.), finishes (smooth, stipple, etc.) and colours (grey, green, etc.) are available. As concrete tiles are usually single lapped, they have to be interlocking, to provide good weathertightness.

Clay tiles have become relatively more expensive because their manufacture does not lend itself to mass production techniques so readily. Indeed, the problem of delamination seems to be associated with machine-made extruded clay tiles.

Clay tiles are more absorbent (up to 10.5% is permissible in BS 402: 1990), and this can result in more vegetation on the roof and a greater risk of frost damage. A higher roof pitch (> 35°) is desirable to prevent this.

Concrete tiles do not delaminate, have a much lower water penetration (see Table 2.1), and so have a better durability record. If they have a coarse finish, however, this can encourage the growth of organic matter such as moss and lichen, particularly in rural areas, or where the roof is near masses of vegetation.

Lime

Lime is produced by burning a fairly pure limestone in a kiln at between 900 and 1200 °C. This lime – strictly speaking, quicklime – is subsequently slaked by the addition of water. Slaked lime forms lime putty, which, with the addition of sand, forms lime mortar (Scottish Lime Centre, 1995). Such lime mortar is known as non-hydraulic, because it will set by reaction with atmospheric carbon dioxide (Appendix B).

Lime plasters were traditionally reinforced with the addition of horsehair. A binder such as tallow or casein, or even blood, would also be added. Lime washes were made from the excess water from the slaking process, and pigments and binders would be added. Even so, regular reapplication of such coatings would be required to maintain a good appearance and protection of external lime-based renders (Boynton, 1980).

One of the great merits of lime-based mortars, renders and plasters is their excellent workability, a property that cannot be wholly conferred on cementitious products. The curing reaction with carbon dioxide in the atmosphere enables re-curing of freshly exposed surfaces, which may develop as a result of thermal or moisture movement (Appendix B).

Hydraulic limes are made from limestones that contain clay. On burning, these constituents combine to form a cement-like clinker, which is either ground to a powder or completely slaked to form quicklime (Wingate, 1990).

Both these forms of lime are used in mortars in a similar fashion to cement. The manner and the rate of hardening and strength development are similar but less consistent and slower than for cement. As well as being a constituent of mortar, lime is therefore an important additive in external render and internal plaster. It is particularly useful, if not essential, in repointing and general building work associated with conservation schemes. Old stonework is likely to have been built using a lime-based mortar. Unfortunately, some insensitive repair programmes involve repointing of stonework in such situations using a cement-based mortar, which is usually far too strong for such work (Scottish Lime Centre, 1995).

Mortar

Mortar is a mixture of sand and other fine aggregates with a matrix or binding agent such as lime or cement. Its purpose is just as much to keep stone and bricks apart as it is to 'bind' them together. While some strength may be necessary in a mortar joint, it must be sufficiently yielding to accommodate movements in the structure. If the mortar is a weaker mix (a lower cement content), the relief of any stresses that develop will be by the formation of a series of fine cracks distributed inconspicuously in the mortar joints (BRE Digest 362). The mortar must, of course, be durable, and in particular, frost resistant, and it must adhere strongly to the masonry units.

The early forms of mortar were lime and sand based. Good working properties were obtained, but hardening was slow, and developed from the surface inwards over a period.

It is current practice to use cement-based mortars either with lime, or more frequently using a plasticiser as a substitute for lime. This form of mortar is much stronger, and with many types of masonry moisture and thermal movements will lead to fractures, particularly at the mortar/masonry joint. If lime is retained in the mix, plasticity is retained and the final strength of the mortar is reduced. The actual amount of lime used will depend on the rate of hardening required. The mortar should be appropriate to the strength of the masonry units. Still, conservation and aesthetic requirements in the context of renovation or restoration will probably necessitate the use of lime-based mortars for such work.

The use of air entrainers in mortar has provided satisfactory workability and good frost resistance. Vinsol plasticisers are especially popular, but accurate gauging of the correct quantity on site is essential if the strength of the mortar is not to be seriously reduced. Current practice in the selection of mortar mixes is included in BS 5628 Part 3 (Fig. 3.4).

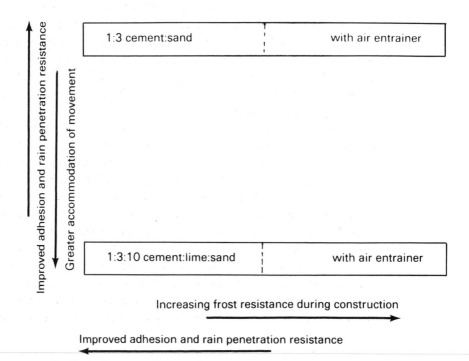

Fig. 3.4   Durability of mortar mixes (BS 5628 Part 3).

If dense impermeable mortar is used with more permeable masonry units – bricks, for example – flow of water during wet weather is likely to take place through the bricks, and any salt crystallisation that may follow on subsequent drying out will take place in the bricks or masonry units, ultimately leading to deterioration. If the mortar strength is lower, drying out and crystallisation will occur in the mortar, which even if it deteriorates can be repaired more easily. Furthermore, cement lime mortars will have a lower salt content, because cement itself produces some salts on curing.

External rendering

Rendering is a cementitious coating applied to external walls. It is very common in Scotland and in other areas of north Britain. Its primary purpose is to provide a weathering protection skin to the brickwork: hence its popularity in regions with inclement weather. It also has an aesthetic function, in that it can give a very attractive finish to walling when used correctly.

Early forms of rendering were based on clay and, as wattle and daub, were employed as weather-resistant coating on 'infill panels' in medieval timber-framed buildings. Stucco is a form of lime and sand rendering, found on the outside of old buildings, which was then painted. Hydraulic limes with more weather-resistant properties were later used in combination with sand. In this form, such products have been used as external rendering up to the present day.

Lack of durability when exposed to frequent wetting and frost action necessitated frequent maintenance, and to this end such renders were often painted with an oil paint or given a coat of limewash. Good performance has been achieved in the more protected situations.

Since the early 1900s Portland cement has been used in renders, and when used in conjunction with lime and good-quality sand will perform well. A mix of $1:1:6$ cement : lime : sand can reduce water content in underlying brickwork by about one third, and $2:\frac{1}{2}:4\frac{1}{2}$ mixtures by as much as a half. Inadequate attention to the selection of materials and the use of incorrect mixes can lead to early local failure in renders, which can actually make conditions worse in underlying masonry. The bare substrate in such situations will be exposed to higher levels of moisture penetration.

The basic requirements for render are that it must adhere to a backing, but should not be too strong, which might lead to detachment. It should have a minimum drying shrinkage, and should resist the passage of water. It should resist erosion, mechanical damage and frost attack (BRE Digest 196).

The main principle for the application of rendering is that it should be at least 20 mm thick in total, and subsequent coats should never be stronger or thicker than the preceding coats. A typical two-coat application, for example, would be: first/scratch coat 12 mm thick in 1 : 1 : 5 cement : lime : sand mortar; second/finishing coat 8 mm thick in 1 : 1 : 6 mix (BS 5262: 1991).

Six main types of rendered finish are in use today. Their characteristics and performance are summarised in Table 3.4. Various types of fixing are used, including metal angle and edge beading, bell-cast pieces, lathing, welded wire mesh and netting, nails, screws and bolts. All metals must be resistant to or protected from corrosion. The type of background, the preparation of the surface and the choice of rendering system are all very important if good performance is to be obtained (Table 3.5).

Table 3.4   Characteristics and performance of renders.

Type of render	Method of application	Performance in service
Roughcast (wet dash) ('harling')	Final coat of wet aggregate thrown onto the wall	Very good
Dry dash (pebble-dash)	Clean aggregate thrown onto wet mortar	Very good
Scraped	Top coat scraped after application	May become dirty
Textured	Top coat worked after application	May become dirty
Smooth	Top coat trowelled after application	May craze and become patchy, especially with too-rich mixes and too-fine sands
Tyrolean	Cement-rich mortar with lightweight aggregate splattered by machine	May become patchy and dirty

When renders are applied, drying may result in shrinkage to the extent that cracks develop in the render itself, and the bond with the backing is also weakened. Differential moisture movement between the render and the backing may cause further weakness to develop in service, particularly if defects allow rainwater to penetrate. Shrinkage will be greater if the sand has not been well graded and contains too many fines with too much water, or if too much cement has been used in the mix (Fig. 3.1). The grading of such sands in the British Standards is intended to reduce the proportion of fine particles (Table 3.6).

Table 3.5   Rendering masonry.

Background	Example	Surface preparation	Comments on render system
Dense and smooth strong materials	Concrete Engineering bricks Dense clay bricks	Poor key Raking joints Bush hammering Abrasive blasting Expanded metal Wire mesh	Stipplecoat or splatterdash for base coat to improve key Render thickness not more than 15 mm thick
Moderately strong and porous materials	Clay, sand and lime bricks Lightweight aggregate concrete blocks	Good key Splatterdash may overcome uneven suction	
Moderately weak and porous materials	Soft bricks Lightweight and aerated concrete	High suction Surface wetting may relieve high suction	Stipple or splatter coat using air-entrained cement Use a water-retaining additive
Crumbling or unsatisfactory backgrounds	Painted surfaces timber	Use metal lathing or expanded metal	Use 3 : 1 sand : cement Three coats, minimum 16 mm cover
Sulphate contaminated		Eliminate moisture Keep dry	Use sulphate-resisting cement

Table 3.6   Grading of building sands (BS 1199, BS 1200: 1976, amended 1984/86).

	Sands for renders and plasters			Sands for mortars	
BS sieve	Type A	Type B	Type M[a]	Type G[b]	Type S
6.30 mm	100	100		100	100
5.00 mm	95–100	95–100		98–100	98–100
2.36 mm	60–100	80–100	65–100	90–100	90–100
1.18 mm	30–100	70–100	45–100	70–100	70–100
600 µm	15–80	55–100	25–80	40–100	40–100
300 µm	5–50	5–75	5–48	5–70	20–90
150 µm	0–15	0–20		0–15	0–25
75µm	Up to 5	Up to 5		0–5(10)	0–8(12)

[a] Type M fine aggregate (BS 882: 1992).
[b] Higher cement content expected.

With pebble-dashing, or dry-dash render, suitable graded stone is thrown onto to a freshly applied cement render. With roughcast or harling, the gravel is incorporated in the wet mix. Any drying cracks that occur, either during initial drying or in service, are more likely to be associated with the pebbles, which tend to throw off water rather than allow it to penetrate into the render and the background masonry.

Unprotected solid walls or brick and blockwork are potentially vulnerable to rain penetration, even though the masonry units themselves are impermeable (BRE Good Building Guides 23 and 24). Where dense impermeable blocks are used, rainwater may concentrate at the mortar joints and penetrate through any fine cracks. More permeable masonry units have a more uniform absorption, so excess water does not accumulate so much at the weak points. Render covers such fine cracks and vulnerable areas, and normally rainwater would not penetrate to cracks in underlying masonry. However, if the render becomes cracked, rainwater is admitted. The presence of the render would retard evaporation of such water, and the dampness might penetrate through the wall.

Surprisingly perhaps, water is less likely to penetrate through absorbent rendering than through cracks that might develop in a strong dense mix. This is because any subsequent evaporation would be slower from a dense but cracked rendered surface.

Rainwater penetrating behind render, especially if the render is dense and impermeable, may cause lack of adhesion, further cracking and subsequent disintegration by frost action. Lack of adhesion causes detachment, and this is manifested in bossed or spalled rendering. This is problematic for a number of reasons: it is unsightly; it exposes the wall to further attack; and falling rendering pieces, if from first floor and above, could be dangerous for passers-by.

Generally, rough rendered surfaces, pebble-dash or dry-dash are preferred, because rain falling on a smooth surface with little absorption tends to fall unevenly in streaks. A rough surface breaks up the rainwater and avoids such concentration of water flow.

Many of the older renders and mortar systems allowed the walls to 'breathe' and thus, during dry weather periods, allowed walls and buildings to lose some of the excess moisture that might have developed as a result of rising or penetrating damp or condensation. Modern dense impervious renders and finishes may actually aggravate damp problems by preventing this drying-out process.

Plaster

Internal plastering provides a continuous and smooth level surface to walls and ceilings and a background suitable for decorative finishes (BS 1191). Harder finishes are more resistant to mechanical damage, whereas a softer, more absorbent finish is to be preferred if intermittent condensation is expected.

Sand, cement, lime and gypsum plaster (Appendix C) based on calcium sulphate are all used for various types of plastering, and different combinations and a range of premixed plasters with lightweight aggregates are available, with special (e.g. acoustic) properties. Lime-based plasters applied in three coats are common in older properties. In such cases they are usually applied to a lath background, consisting of wooded laths approximately 35 mm wide × 3 mm thick fixed to 25 × 20 mm timber battens at 300 mm centres fixed to the masonry. Such a finish is now used only in repair or restoration work in historic buildings. It has been superseded by cement- and gypsum-based types, with their faster rate of hardening.

It can be difficult at face value to determine the source of a dampness problem on a lath and plaster wall without cutting out some of the finish to expose the substrate. Care must therefore be adopted when taking WME readings on lath and plaster walls.

Both lime and gypsum plasters may need reinforcement in certain use. Human and animal hair, chopped straw and reed, hemp and jute have all been used In the past. It should be borne in mind that these organic substances will, of course, provide another source of food when affected by dampness-related problems such as fungal attack and mould growth.

Lightweight supports, in the form of hazel twigs interwoven around willow or oak slats, often cleft, were in time replaced by the more familiar sawn softwood laths referred to above. Expanded metal and plastic lathings are now more commonly used. These new types of lath are essential in plasterwork on backgrounds consisting of two or more different types of material (e.g. a concrete lintel in a brick wall), to accommodate their likely differential moisture and thermal movements. They can also be used to provide a mechanical key for plastering on non-absorbent backgrounds.

The development of lightweight plasters incorporating aggregates such as expanded perlite or exfoliated vermiculite offers better insulation properties, and they are less prone to condensation (BS 1191: Part 2). Carlite browning is a common example of this type of plaster, but it is quite unsuitable for use in situations where dampness has occurred or is anticipated.

When older, less alkaline, plaster becomes damp as a result of condensation, mould growth is a common development. This reduction in alkalinity is caused by natural carbonation of the masonry over time. The high alkaline surface of new cement and lime-base plasters seems to discourage the growth of such moulds.

Stucco is a form of plaster used externally in older buildings.

Plasterboard

In the manufacture of plasterboard, a core of gypsum plaster is enclosed between and firmly bonded to two sheets of heavy paper in a continuous process (BS 1230: 1994). The paper that is selected may have a rougher texture (and grey face) if a finish (skim) coat of plaster is required or a smoother one (and ivory face) if a self-finish with a wallpaper is desired.

Plasterboard is also manufactured with either an aluminium foil or polythene backing (British Gypsum, 1991). In the case of foil-backed plasterboard, the bright metallic surface reflects heat, thus improving insulation, and can provide some protection against water penetration. Polythene-backed plasterboard is used where a degree of vapour control is required from the finish.

Plasterboard with foam plastic, phenolic, polyurethane, polyisocyanurate or expanded polystyrene backing (usually about 25–35 mm thick) is also available for use where improved 'internal dry lining' insulation is required. All these products give off a lot of smoke when burnt, and polyurethane gives off HCN as well. Phenolic foam insulation performs relatively well in fire conditions. It does not give off as much toxic smoke as substances such as polyisocyanurate, which gives off cyanide gas when burnt.

Plasterboard should never be used in situations where persistently high-humidity conditions or dampness can be expected (British Gypsum, 1991). Its general use, however, is very popular because of the move away from 'wet-trade' finishes such as plastering. This in turn reduces the amount of construction moisture generated and thus helps the building to dry out quicker.

Sealants

The function of a sealant is to prevent the ingress of moisture, draughts or dirt, either alone or in association with other joint elements. It must be able to perform this function through changing environmental conditions that might cause thermal or moisture movement and, throughout, adhesion must be obtained with the joint surface (Beech, 1981; PSA, 1989).

One of the oldest-established sealants used in the building industry is putty, which is made by intimately mixing a whiting, which is a form of ground chalk, in linseed oil. Linseed oil putties skin overnight but need four to eight weeks to set firm. The linseed oil oxidises, and the putty dries by excess oil being absorbed if the substrate is of wood. Priming the surface of the 'weather rebate' on the timber frame should ensure the correct rate of absorption of the excess oil. Linseed oil putty has fair adhesion but has a low life expectancy, which is dependent on the protection provided by the covering paint film. Solvent-based acrylic sealants are often recommended for use on wood that has been coated with wood stains, where conventional linseed oil putty would dry too quickly and soon lose plasticity.

Ordinary linseed oil putty is not suitable for use on metal frames. Putty for use in metal frames needs to have a lead additive to provide a harder set: hence the grey colour of such putties.

Mastics are another popular form of sealant, and they can be applied by extrusion through a mastic gun, by trowelling, by pouring, or as tapes or strips. They should be easy to apply and should retain their form, adhere well to the sides of the joint, accommodate movement and be permanent. Priming the surfaces to which mastic is to be applied has been found to be beneficial in improving the long-term performance of the mastic. Mastics that are oil or bituminous-based have a limited durability, perhaps up to 10 years, and have limited ability to recover after deformation of over 10% of the original size. They form a tough skin on oxidation of the surface; the material inside remains plastic, but this progressively hardens, leading to ultimate failure. They are mainly used for sealing gaps around window and door frames.

Polysulphide, polyurethane and silicone sealants are generally more expensive. They tolerate greater movement in the joint, and have a greater anticipated service life of up to 20 years. They set by chemical curing. The degree of exposure, and especially temperature fluctuations, has a great effect on performance.

Sealants usually fail in one of four ways:

❏ adhesion failure;
❏ cohesion failure;
❏ fatigue failure;
❏ slump failure.

These problems can be prevented by using the correct type of sealant in the appropriate joint design (Fig. 3.5).

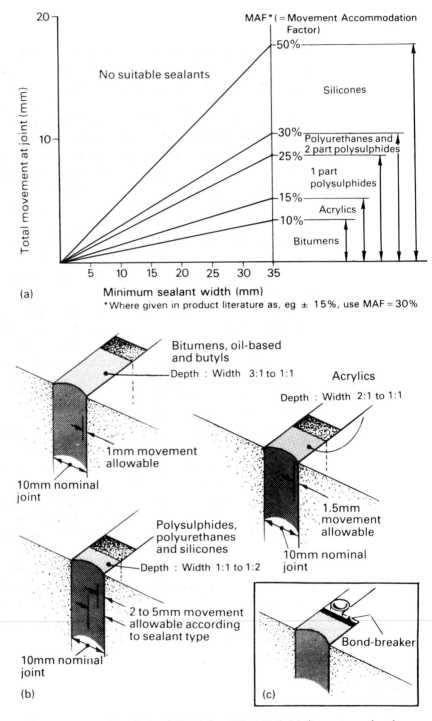

Fig. 3.5 Principles of joint design (after PSA, 1989): (a) dimensions of sealants; (b) back-up materials; (c) bond-breakers.

Asphalt

Asphalts are derived from two sources: natural rock asphalts and lake asphalts (BS 6577). Natural rock asphalt consists of a limestone that is impregnated with bitumen, and it occurs in various geological formations in Central Europe. Lake asphalts come from lake deposits, the main source being Trinidad. These contain higher proportions of bitumen, and up to 30% clay. To produce a useable building material, both products have to be heated, mixed and blended to achieve a bitumen content of between 11 and 14%. A proportion of finely divided material is necessary to assist flow during working operations, and coarser gritty sand is added to give the material 'body', forming 'mastic' asphalt.

When delivered to site, mastic asphalt arrives in the form of 'cakes' about the size of concrete building blocks, which are gradually heated to reduce the mastic to a workable state. Overheating to above 215 °C can be harmful. Overworking and overtooling of the asphalt can lead to early deterioration (BRE Digest 144).

Asphalt is a traditional method of covering flat roofs. Properly applied it can provide a service life of 20 years or more. However, precautions must be taken to prevent premature deterioration of the asphalt, primarily through either incompatibility with other materials or solar degradation. Asphalt has a different coefficient of thermal expansion from the other materials it is likely to come into contact with (e.g. timber and concrete deckings). It is also chemically incompatible with materials such as plastics, which may suffer from plasticiser migration if in direct contact with asphalt. This problem can be avoided by using an inert fleecing layer between the asphalt and the substrate.

Solar radiation can cause blistering and even melting of the asphalt. This problem can be minimised by applying two coats of a metallic-based aluminium paint to the finished asphalt. Such solar protection, however, erodes over time, and thus may require renewal every five years or so.

Asphalt can also be used as a damp-proof membrane in floors, and as a tanking system to prevent damp penetration in below-ground situations. One of the main advantages of this 'poured' tanking material is its flexibility. It can thus be used in awkward-shaped areas more easily than can sheet materials (see Chapter 8).

Bitumen

Bitumen is a residue obtained either from the distillation of crude oil or in natural deposits. In several building products it is used in a water-based emulsion or as an organic solvent-thinned material. It is used both as a waterproofing material and to prevent or retard corrosion. It has good acid- and salt-resisting properties. It has useful adhesive properties, and bituminised building paper has widespread uses as a water and vapour barrier. It is very susceptible to organic solvents, and so special care has to be taken when using such solvents in proximity to bitumen materials.

Bituminous felts consist of a reinforced base coated with bitumen, and two or more layers of felt, applied on site, form built-up felt roofing (see Chapter 5). The bitumen provides the waterproofing and adhesive properties, and the nature of the base will determine the strength, weathering and ageing characteristics of the felt.

Early forms of bitumen roofing felt used rag, wool or asbestos for the fibre base, but in the 1950s glass fibre tissue reinforcement was introduced. Also, more recently polymers have been added that modify the bitumen to make it more flexible in winter, firmer in summer and more fatigue resistant. Polyester felts, however, are stronger. In time, bitumens crack and craze as a result of ultraviolet light, oxygen and ozone, and this results in increasing hardness after 15 years' exposure. Styrene butadiene styrene (SBS) and atactic polypropylene (APT) additives improve flexibility, strength and fatigue resistances in bitumen. The use of asphalt and bitumen for flat roofing is described further in Chapter 5 and for tanking basements in Chapter 8.

Metals

Dampness and its effect on metals

Metals are used widely in both old and modern buildings. They are used in old properties for rainwater goods, fixings, ironmongery and railings. In modern buildings they are more used for structural components, claddings, electrical installations, internal plumbing and heating, as well as for ironmongery and fixings.

In nature most metals are found in the form of ores, in which the metal is combined with either oxygen, carbon or sulphur. This is because metals are reactive materials. They have to be separated from these other elements before they can be used, and then most of them have to be protected from the environment in order to prevent them from reacting, or corroding.

Wet or aqueous corrosion can occur when metals are in contact with other damp materials or in humid atmospheres. The presence of dust, dirt or corrosion products (e.g. salts) on metal surfaces, combined with the high thermal conductivity of such a material, can give rise to condensation at humidities as low as 60% RH. In other situations, adjacent building materials may act as sponges, wicks or water traps (i.e. reservoirs), and thus increase the corrosion risks.

The corrosion of metals and alloys is essentially a process in which some of the metal is removed and goes into solution, where it may remain, or may precipitate out on the metal surface, like rust on steel. It is an electrochemical action in which different metals corrode at different rates.

The tendency for a metal to react with the environment and corrode is expressed in the electrochemical series (Table 3.7). Metals such as sodium will react violently with cold water and corrode away in a few minutes, whereas gold will remain unchanged for a long time. Most common metals lie between these two extremes.

Table 3.7   Electrochemical series.

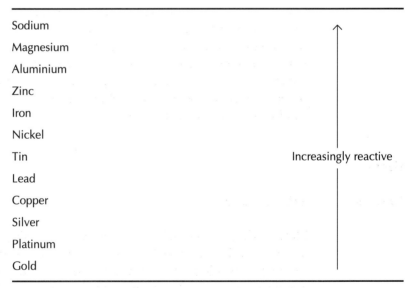

Sodium	
Magnesium	
Aluminium	
Zinc	
Iron	
Nickel	
Tin	Increasingly reactive
Lead	
Copper	
Silver	
Platinum	
Gold	

Note: The more widely separated the metals are in this table, the more rapid and greater the effect of galvanic corrosion.

In some cases, galvanic corrosion can be used in the form of sacrificial protection. Zinc galvanising of iron can provide protection to the iron in this way.

Galvanic corrosion

When two different metals are electrically connected together in a damp situation, the corrosion effects are often greater. This is because an electrical potential (voltage) is generated, and the more widely separated the two metals are in the electrochemical series, the greater the reaction will be. If copper and aluminium are connected and placed in solution, the aluminium will corrode while the copper is protected. The more susceptible metal is said to be the anode, and the less susceptible the cathode. This is known as galvanic (or bi-metallic) corrosion, or electrolytic action. Water is a primary electrolyte in corrosion of metals.

Thus corrosion of cast-iron tanks connected to copper piping in a domestic water system can be expected at the junction. If the area of the cathode is large in relation to the anode, then the corrosion of the anode can be very severe. An iron nail in a copper sheet will corrode very rapidly, whereas the corrosion of a copper nail in a steel sheet would be less serious. Therefore certain combinations of different metals, which can arise from bolting riveting, welding, braising and soldering, are to be avoided (Fig. 3.6). Where this is impossible, the metals must be insulated from each other by non-conducting and water-resistant washers and tapes. Neoprene rubber gaskets have proved very successful in such situations.

	Metals	1	2	3	4	5	6	7	8
1	Copper								
2	Phosphor bronze								
3	Aluminium bronze								
4	Stainless steel								
5	Mild steel								
6	Manganese bronze								
7	Aluminium								
8	Cast iron								

Fig. 3.6 Recommendations for bimetallic contacts (from Harris & Edgar, 1995): these metals can be used together in all conditions; these metals can be used together in dry conditions only; these metals must *not* be used together. *Note:* dry conditions, i.e. cast in or within a cavity above dpc level, except where the cavity is used for free drainage. Marine conditions require special consideration.

Passivity

The simple picture of corrosion susceptibility presented by the electrochemical series is modified by the development of protective oxide films on metal surfaces. In some cases these oxide films are so soluble that corrosion can proceed unhindered. The formation of aluminium oxide on the surface of aluminium is a familiar example of the way in which an oxide film can protect a metal from corrosion. Some alloys have been designed to have a protective oxide film. Thus steel is made stainless by the addition of 12% chromium, which is enough to alter the nature of the iron oxide film and stabilise it.

If an oxide film is to be protective it must have certain properties. It must:

❏ be impermeable;
❏ have a strong bond to the base metal;
❏ resist chemical attack;
❏ have good mechanical strength;
❏ have a self-repairing ability.

The film formed on aluminium has these qualities, making the metal passive in ordinary water. The film of iron, by contrast, does not form a passive layer in ordinary water, so corrosion 'rusting' takes place. Passive layers may lose their properties if acidity or alkalinity changes. For example, aluminium can corrode if it is placed in or on concrete. If salts such as chlorides and sulphates are present, local breakdown of the passive film may occur, and pitting may develop.

Corrosion of lead

Generally lead is very resistant to corrosion in normal exposure conditions. A film of lead carbonate or sulphate, called a 'patina', helps to protect it from the effects of atmosphere and rainwater. Corrosion of lead, however, can occur as a result of:

❏ condensation (see Chapter 6);
❏ exposure to acidic conditions (e.g. acid rain, acetic acid fumes, and acids leached by timbers such as oak and organic growths such as algae, moss and lichen);
❏ exposure to alkalis, especially under damp conditions (e.g. when in contact with lime or cement mortars);
❏ electrolytic action (e.g. when in contact with a dissimilar metal).

Corrosion of alloys

The corrosion of alloys may be more complex that that of pure metals. As alloys contain more than one metal component, the more active will tend to dissolve out at a faster rate. This can cause local pitting or selective deterioration of the alloy.

Environmental influences on corrosion

Increased corrosion rates can be expected where atmospheric pollution with sulphur dioxide occurs (from coal/oil-fired emissions). Also, salts deriving from marine conditions, building materials, and the ground can have similar effects. In soils, some bacteria can reduce sulphates and thus cause severe attack on iron pipes.

Corrosion of metals in association with timber

Most basic metals and alloys can safely be used in contact with dry timber. At moisture contents below 14 to 16%, little corrosion occurs.

The majority of timbers are acidic, with an average pH value of 5 (Coggins, 1991). Western red cedar and Douglas fir, for example, are markedly so. The more resistant metals and alloys such as copper, aluminium, brass, bronze and stainless steels should be used in conjunction with timber. If less resistant materials and alloys are used, such as steels, they should be protected by a coating of zinc galvanising. The corrosion of steel in timber can also cause local damage to the timber. The tensile strength may be reduced, and localised softening may reduce or weaken the holding power of the fastenings. Some water-based preservatives and fire-retardant treatments (containing organic salts) may increase corrosion if the treated timber becomes damp (BRE Digest 301).

Methods of metal protection

Provided that metals remain dry, corrosion will not develop. If this is not possible, more resistant metals and alloys should be selected in place of steel. Alternatively, coatings of plastics, metallic paints, micaceous oxide paints, microporous paints, rubbers or enamels can be used to prevent the metal from coming into contact with the environment. To be effective, coatings must be applied to carefully prepared surfaces; they must remain intact, and may require regular maintenance. Chemical coatings such as phosphating of steels are usually applied prior to painting; on aluminium alloys, anodising may be the only treatment required.

Chapter 4
Dampness and Wood

Moisture in wood

Wood and water have a very special relationship with each other. This is due in part to the fact that water is essential for the growth of trees, and is indeed the basis of the process by which all plant cells are formed and plant tissues initially achieve rigidity. As a result, wood, like many other building materials, has to be dried if it is to perform satisfactorily when used in buildings. The drying process causes many changes in wood properties that are unique and quite different from those of other materials. This is probably one of the main reasons why timber is so often misunderstood and misused by the building industry.

When examining the cross-section of the trunk of a tree (Fig. 4.1), inside the bark proper can often be seen to be clearly divided into two zones: the outer part, known as the sapwood, and the heartwood in the core. Sapwood represents perhaps 25% of the cross-sectional area of the trunk, and in the living tree it is the zone through which the moisture (sap) is conducted from the ground to the leaves.

In a standing tree, the sapwood may have a moisture (sap) content of 100 to 200%. As the tree increases in girth, the sapwood is converted into heartwood, where the conduction of the sap ceases. A much lower moisture content (about 50%) in the heartwood in the tree is therefore likely. In some trees heartwood is also darker in colour and more resistant to fungal decay.

When examined under the microscope (Fig. 4.2), wood can be seen to have a cellular structure. The passage of water in the tree is achieved through a system of hollow interlocking fibre-shaped cells that provide the channels for this water flow. As would be expected, the flow is much greater along the axis of the tree, and the line of these cells constitutes the 'grain direction' of the timber. Dry wood consists of a cellular structure of these wood cell walls (Fig. 4.2).

Wood cell walls are made up of two natural polymers: one is cellulose,

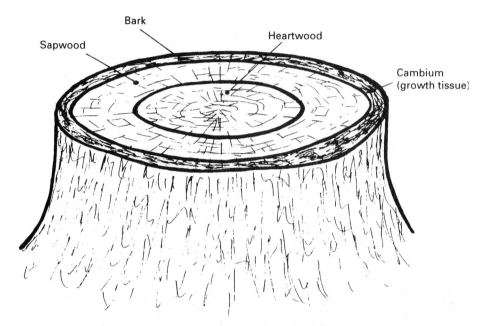

Fig. 4.1   Cross-section through the trunk of a tree to show the different parts.

which is present in the form of long chain molecules embedded in a matrix of the other polymer, lignin. The structure of the wood cell wall can be likened to reinforced concrete, in which the cellulose chains (steel reinforcement) provide the tension strength along the axis or grain of the wood; the lignin (concrete) surrounds the cellulose (reinforcement) and protects it.

The water in wood is present in two forms: free water, in the cell cavities, removed first during drying; and combined water, present in combination with cellulose in the cell walls, even in relatively dry wood such as in buildings or furniture. The moisture content of timber is expressed as a percentage of the oven dry weight:

$$\text{Timber MC (\%)} = \frac{\text{Initial weight} - \text{Oven dry weight}}{\text{Oven dry weight}} \times 100$$

In green timber, the moisture content of sapwood might be over 150%. When dried, either naturally (i.e. by air seasoning) or artificially (i.e. by kiln drying), the water in the cell cavities (the 'free water') is lost first (Fig. 4.3).

When the timber is dried to a moisture content of about 25–30% – known as fibre saturation point – the majority of free water has evaporated, and further drying will cause the water combined with the cellulose in the cell walls to dry out. As this takes place, certain changes

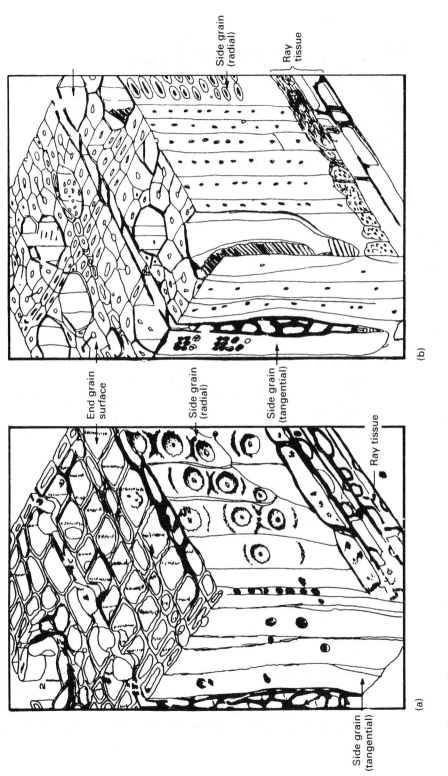

Fig. 4.2   The structure of (a) a softwood and (b) a hardwood (× 225 approx.).

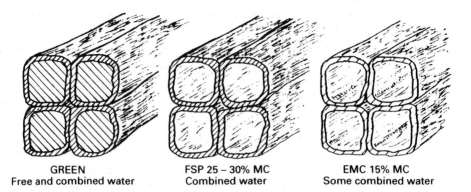

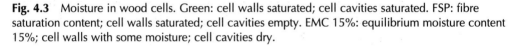

GREEN	FSP 25 – 30% MC	EMC 15% MC
Free and combined water	Combined water	Some combined water

Fig. 4.3 Moisture in wood cells. Green: cell walls saturated; cell cavities saturated. FSP: fibre saturation content; cell walls saturated; cell cavities empty. EMC 15%: equilibrium moisture content 15%; cell walls with some moisture; cell cavities dry.

in the wood properties follow: in particular, strength properties increase, susceptibility to fungal decay decreases, penetration of wood preservative generally increases, painting and gluing is more successful, electrical conductivity decreases, and movement (shrinkage) occurs as the dimensions are reduced.

All these changes are reversible if the wood increases in moisture again. This is because the cellulose in the cell walls of the wood has a special relationship, in fact an attraction, to moisture in either liquid or vapour form. This is described as hygroscopicity. The cellulose in the wood is hygroscopic, and is always tending towards a moisture content in balance with the moisture in the surrounding environment: the so-called equilibrium moisture content (Table 4.1).

Table 4.1   Equilibrium moisture contents of timber in buildings.

Location	Type of joinery	EMC (%)
External joinery	Fascias Bargeboards Windows/doors	16–18

Internal joinery	Structural and joinery timber:
	  with intermittent heating
	  with central heating
		12–14
		8–10

In a correctly designed and maintained building, liquid water should be excluded from the fabric. Provided this is achieved, the principal cause of fluctuating moisture content will be variations in relative humidity. Of course, humidity levels are always changing, and so the timber moisture content will be in broad balance with the average environmental humidity.

The desired (i.e. safe) moisture contents of timber in service are often referred to as equilibrium moisture contents, or EMC (Table 4.1). If wood is dried to a moisture content that is in equilibrium with the level of humidity of the environment, and if these conditions do not alter significantly, the problems of changing dimensions (movement) are avoided. In practice, these ideal conditions of stable humidity and total exclusion of water do not always arise, and movement frequently occurs.

The small amount of wood movement along the grain (lengthwise) can be ignored for all practical purposes. However, lateral movement is greater in the tangential direction (i.e. tangential to the growth rings) than in the radial direction (i.e. in the direction of the rays). Movement averages about 0.275% tangentially for each percentage moisture content change and 0.125% radially (Fig. 4.4). The figures quoted are for Baltic redwood (*Pinus sylvestris*), which is considered to be a timber with a 'medium movement'. Other timbers with greater or lesser movement are described as large or small movement species respectively (Table 4.2).

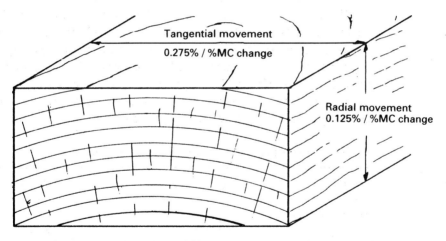

Fig. 4.4 Lateral movement of wood caused by moisture content change below 30%. Movement along the grain can be ignored. (The figures quoted are for Baltic redwood.)

As mentioned earlier, the cellulose in wood cells is hygroscopic, and this results in the water becoming 'attached' to the cellulose during periods of moisture increase and being lost by evaporation during drying. As the cellulose is present in the form of long chains in the wood cell walls, and these chains are orientated in the 'grain' direction, we could imagine water molecules pushing their way into the cell wall and forcing

Table 4.2   Movement classification of timber.

Degree of movement		
Small movement (less than 3%)	Medium movement (3–4.5%)	Large movement (greater than 4.5%)
Afrormosia	Keruing	Beech
Douglas fir	Oak	Birch
Western Hemlock	Scots pine	Ramin
Iroko	Sapele	
Larch	Walnut	
Mahogany		
Seraya		
Teak		
Western red cedar		
Whitewood		

these cellulose chains apart, causing the wood to swell laterally. Conversely, loss of these water molecules causes shrinkage in the lateral dimensions as the chains draw together again. We could further suggest that the lower radial movement is because the rays restrain movement in that direction. Although not all research supports these theories, they represent a simple explanation of the facts.

If wood is in contact with liquid water in a building, the pick-up will be much more rapid along the grain because of the cellular structure. The permeability from the end grain is many hundreds of times greater than from the side grain surface.

As timber dries, most of its strength properties increase. Below fibre saturation point (25–30% moisture content), modulus of elasticity (i.e. the measure of a material's stiffness or flexibility) shows an increase of 2% for every 1% moisture reduction. Drying also leads to an increase in the ease of penetration of liquids – wood preservatives, for example – as the loss of free water leaves empty cell spaces for penetrating liquids.

Wood-based products

Softwoods and hardwoods are not the only wood-based products used in building nowadays. Composite wood-based products, either in sheet or strip form, are being used to a large extent as a substitute for these 'natural' timbers. These products were traditionally not good at resisting moisture. However, some composite wood-based products made with

Table 4.3 Typical wood-based building materials.

Material	Binder	Uses	Remarks
Chipboard BS 5669	Heat and pressure using a resin	Floor and roof decking; worktops (see BRE Digest 373)	Not suitable for use in cold-deck roofs
Wood-fibre board BS 1142: 1989	Heat and pressure	Insulation	Not suitable for use in damp conditions
Oriented strand board (OSB) (BRE IP 5/86)	Ditto	Panelling; sarking; shuttering; flooring	Not suitable for use in high-humidity conditions
Plywood (BS 6566: 1985)	Water- and boil-proof (WBP) or moisture resistant (MR)	Roof decking; panelling; ducting; sheathing	Only WBP ply must be used for external conditions
Medium/high-density fibreboard (BS 1142: 1989)	Compression	Linings; facings; skirtings; claddings; mouldings	Not suitable for use in damp conditions
Cement-based particleboard BP IP 14/92	Cement and compression	Decking; cladding; flooring	Not suitable for use in cold-deck roofs
Hardboard BS 1142: 1989	Compression and heat	Panelling; floor sheating	Not suitable for use in damp conditions
Woodwool slabs (BS 1105: 1994)	Cement and compression	Flat roof deck; permanent shuttering to concrete walls and soffits	Not suitable for use in high-humidity conditions

waterproof glues can be used in exterior conditions. The main examples of wood-based products used in buildings are shown in Table 4.3.

Fungal decay

Wood, like humans, needs water to survive. Too much water, however, can be deleterious to wood. It can give rise to a whole range of fungal-

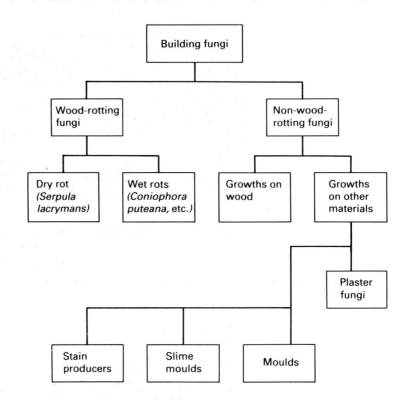

Fig. 4.5   Types of fungus in buildings.

related problems (Anon, 1986). The main types of wood-rotting and non-wood-rotting fungus are summarised in Fig. 4.5.

High timber moisture content leads to susceptibility to decay. Above 30%, wet rot can develop, and the optimum for this type of decay is around 50–60%. Dry rot, on the other hand, can cause decay in timber at a moisture content as low as 20%, though most rapid decay generally develops at 30–40%.

The actual processes of decay are initiated by the growth of microscopic root-like fungal hyphae (strands) which grow in the wood (Figs 4.6, 4.7) and which liberate digestive chemicals (enzymes) capable of breaking down the polymers that form the wood cells into their constituent parts. Thus cellulose is broken down into glucose, and ultimately to water and carbon dioxide. The actual degradation of wood cellulose into glucose by wood-destroying fungi, while occurring only at high moisture contents, actually produces still more moisture when glucose is ultimately converted to carbon dioxide, and theoretically the water produced equals over half (55.5%) of the cellulose decayed by the fungus.

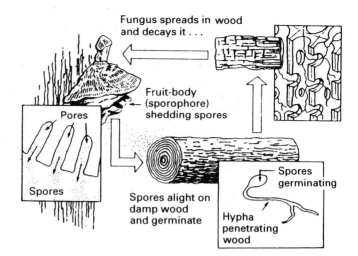

Fig. 4.6   Life-cycle of wood-destroying fungus (after TRADA).

Fig. 4.7   Fungal hyphae in wood (after TRADA).

There are basically two groups of timber rot: brown rots and white rots, which relate to the colour of the wood in its decayed state. The two main brown rots are *Serpula lacrymans* (the true dry rot) and *Coniophora puteana* (a wet rot). All white rots are forms of wet rot (Richardson, 1993).

Dry rot

Background

The true dry rot fungus *Serpula lacrymans* (formerly called *Merulius lacrymans*) is the most destructive fungus causing decay of timber in buildings (BRE Digest 299). It derives its Latin name from the words *serpula* (meaning creeping, as in serpent) and *lacrymans* (meaning tears, as in weeping). It produces specialised conducting strands that support growth across timber and into damp masonry where hidden timbers – bond or fixing grounds, for instance – are likely to be affected, as well as adjacent joinery. The role of these strands has frequently been misunderstood in the past, but recent research suggests that they are formed behind the advancing 'front' hyphae (Coggins, 1991). The hyphae spread into any area where the ambient humidity is high enough and where conditions of both humidity and temperature are stable and not subject to rapid change. The strands are formed 'in the rear' of the 'front', and conduct nutrient materials from the food base some distance away. They may also assist in the spread of the fungus by conducting moisture to increase humidity and moisture conditions, but only under conditions where impervious materials like polythene, plaster or render restrict ventilation and the drying of the timber. Thus dry rot may 'wet up' conditions that are otherwise marginal for growth. When dry rot is found growing under high humidities, water droplets can often be seen on the mycelium, and the fungus is said to be 'weeping' (lacrymans) (Ridout, 1994).

The concept of 'static' conditions, probably at above 90% RH, describes the situation where moisture generation by dry rot becomes significant. Thus 'dynamic conditions' describes conditions where environmental conditions are too inconsistent in terms of ventilation, temperature and ambient humidity for the additional moisture from the decay process to be significant.

Constantly damp, humid and warm conditions at or above 90% RH and around 23 °C favour dry rot. These conditions can develop in buildings in a number of ways: for instance, in wood panelling or behind dry lining fixed on timber studs adjacent to damp external masonry; under suspended timber floors where sub-floor ventilation is blocked, or inadequate; on timber embedded in persistently damp masonry, either as joist ends, rafter feet, wall plates, or bond or framing timbers.

Thick masonry walls that become damp can take a very long time to dry out, even when the defects that have caused the damp have been identified and cured. It is often estimated that masonry that is saturated will dry out at a rate of 25 mm (1 in) thickness per month, so a 225 mm (9 in) solid wall may take nearly a year to dry out naturally. Some observers suggest that in modern buildings drying out may be more prolonged, because there is less 'natural' ventilation, and more vapour-resistant materials are employed. Thus the conditions favouring dry rot may still exist during such extended drying-out periods. That is why rapid (artificial) drying out may be required; but this must be done carefully to avoid problems associated with excessive shrinkage (see BRE Digest 163).

Dry rot is sensitive to high temperatures, and is killed by exposure to temperatures of over 40 °C for about 15 min (Coggins, 1991). It is therefore anticipated that in typical roof situations dry rot will not usually develop rapidly, because dynamic conditions will prevail with varying temperature and moisture levels. Only where there is adjacent damp masonry – parapet walls, for example – or in the core of large-dimension timbers (such as 'bressummers') would there be sufficient reservoirs of dampness and the moderation of temperature fluctuation necessary for the static conditions preferred by dry rot.

Characteristics of dry rot decay

Active growth of dry rot is characterised by the development of a mass of hyphae called mycelium, which looks like cottonwool (Bravery & Carey, 1993). This mycelium is usually white in colour, sometimes with yellow and purple patches, and under humid conditions is often covered with water droplets (hence the term *lacrymans*). Under drier conditions it collapses to a thin grey-brown skin.

Very frequently this fungus mycelium growth takes place in concealed places, such as behind dado wall panelling, behind 'beam-filling' at the eaves in roof voids, or under floors, and produces few signs of its presence until it has spread some distance and has seriously decayed important structural timbers. Its growth rate is normally about 1 m a year, but if the conditions are ripe annual growth of up to 4 m may be experienced (Hollis, 1991).

The decay turns the wood brown, and it typically cracks into large (up to 50 mm) cubes. Dry rot is described as a brown rot, and with such decay only the cellulose is decomposed, while the lignin remains comparatively unaffected.

Under favourable conditions, or sometimes as a result of stress caused by ineffective or incomplete remedial measures, the fungus will form characteristic fruiting bodies called sporophores. These take the form of flat, mushroom-like plates or brackets, up to 1 m in size. They may be formed on the wood or on the surrounding masonry, and are often the initial visible sign of hidden decay. Sometimes the first non-visual indication of a dry rot attack is its pungent mushroom smell. That is why 'rot hounds' are said to prove effective at tracing an outbreak of dry rot (Hutton & Singh, 1989).

The centre of the fruiting body is the area where the spores are produced. Spores are liberated in vast numbers: millions per hour from a mature fruiting body. The surface from which they produce is known as the hymenium. It is warty and rough, and occasionally covered with spikes or pines. This hymenium is usually dark red-brown in colour, and is surrounded by a thinner white edge, which is the growing margin and which will later become a fertile hymenium layer producing more spores.

The spores are individually very small – each about a hundredth of a millimetre long – but when they accumulate in quantity, near a fruiting body, for instance, they may form a characteristic red dust. These tiny spores can be carried considerable distances in air currents and, if conditions are suitable, any spore that is deposited near timber will germinate and another decay outbreak may be initiated. Spores can remain viable for several years.

Although dry rot needs any material with cellulose (such as timber, paper, cardboard, hardboard, hessian, etc.), the importance of calcium in its development should not be overlooked. During the growth of dry rot the fungus exudes an organic acid called oxalic acid (Bech-Anderson, 1991). This reduces the pH value of the wood from about 5 to 3 (Coggins, 1991). The wood thus becomes more acidic, and if this is not curbed the growth of the rot will cease. Unfortunately, the acid is neutralised by the calcium in the surrounding masonry, and this allows the rot to spread, even through walling. Calcium for example, is found in mortar, bricks, and mineral wool insulation. This, along with a deficiency in the conditions required, may explain why dry rot has never yet been found in an all-timber house (Douglas & Singh, 1995).

Wet rot

While dry rot is economically the most important of the wood decay fungi that are found on damp building timber, there are a number of other species of fungi that cause what we call 'wet rot' (BRE Digest 345). The term is appropriate because, in the main, these fungi require higher moisture levels to cause decay in the timber. The minimum level for wet rot is generally considered to be 30% WME, and the optimum is between 50 and 60%. There are several different species.

Coniophora puteana or Cellar fungus (previously called *Coniophora cerebella*) is probably the commonest wet rot decay of softwood timber. It is a brown rot, but the timber typically cracks into small cubes (i.e. the cuboidal cracking is less extensive than with dry rot), which are sometimes formed internally under a very thin veneer of sound wood on the surface. The most characteristic feature of this fungus is the development of brown branching strands on timber, and occasionally on adjacent masonry. Unlike dry rot, wet rot does not require the presence of calcium for growth (Bech-Anderson, 1991). It rarely forms extensive fruiting bodies on timber in buildings, although if present they are usually less distinctive than the sporophores of dry rot. It is a common cause of decay of ground-floor timbers with poorly ventilated sub-floor conditions or on damp floors under linoleum, and of external joinery.

Fibrioporia vaillantii (Poria) is another common wet rot decay of softwood building timber frequently found on skirtings close to damp walls and on wet roofing timbers as well as on very damp floors. It is another brown rot, but the wood is lighter in colour and the cracks are not so deep as those caused by *Serpula lacrymans* (Bravery, 1991). The most characteristic feature of *Fibrioporia* is the formation of superficial clumps of surface mycelium, white branching strands and occasionally the formation of flat platelike fruiting bodies on timber, on which minute spores can be seen with a hand lens.

There are several other species of fungi causing wet rot of building timber that produce a white rot (Bravery, 1991). In this type of decay, the fungus produces enzymes that can cause the decomposition of both the cellulose and the lignin. In the process the wood turns an off-white/light grey colour: hence the classification. The decayed wood has a soft springy texture, and does not develop the cuboidal cracking that is so characteristic of brown rots.

The characteristics of the three most common species of brown rot found on building timbers are described in Table 4.4.

Non-wood-rotting fungi

Stains

Stain fungi grow on and in timber without causing actual structural weakening or decay. They develop on 'green' timber, both in the log and after conversion. They can continue to grow and cause discoloration

Table 4.4 Identification of some common species of fungus found on building timber.

Name	Effect on wood	Surface mycelium	Fruit body aporophore	Remarks
Serpula lacrymans (merulius) Dry rot	Brown rot	When moist, cottonwool-like masses	White margin, rusty red centre. Up to 1 m across	Masonry and brickwork under damp and humid conditions are penetrated. Found in buildings, boats, and in mines
	Large cuboidal cracking	When dry, silver-grey skin. Yellow-purple and lilac patches	Red-ochre spore dust. Form as a plate or bracket	
		Strands up to 6 mm wide, brittle when dry		
Coniophora puteana (cerebella) Cellar fungus Wet rot	Brown rot Splits along the grain and in small cubes. Often leaves a thin skin of sound wood	Scanty Forms brown branching strands on wood and on masonry	Small flat plate-like, on wood surface. Rare. Yellow margin, olive-green centre. Rare	In buildings and on timber externally. Commonest form of wet rot
Fibroporia vaillantii Pore fungus Wet rot	Brown rot. Cuboidal cracking	White skin and branching strands, which remain flexible when dry	White plate-like growth with pores	In buildings, on mining timbers and externally
Sap stain (blue)	Blue-grey, sometimes brown discoloration throughout sapwood	Black surface growths sometimes develop		Occurs on freshly felled and on converted timber. Ceases to develop once moisture content drops below 25%, but disoloration remains
Mould growth	Various colours of entirely superficial growth	Black, green, white and other colours usually in patches		Occurs on damp timber and under conditions of high humidity. Once dry conditions are established, it will cease to grow, and the discoloration can easily be removed with a light planing or sanding

through the sapwood until the moisture content drops below 30%. The discoloration is usually described as 'bluestain', which has been coined because the growth of the fungus causes a blue/grey staining of the sapwood. However, sometimes the colour can be brown or black, and so 'sap stain' is a more appropriate description.

This discoloration is not normally significant in structural building timber, as it is not associated with any strength loss in the material. Furthermore, the fungus that causes the staining cannot survive in wood below 30% moisture content and will therefore be dead and inactive. Nevertheless, any fungal growth on a timber or building material is an indicator of high moisture levels, and should be investigated accordingly.

Moulds

Mould growth will occur on timber under conditions of high humidity, usually over about 70% RH. Various species of mould can occur, and they can usually be seen as causing small blotches on the surface. Green and white colorations are the commonest, although other mould fungi may cause other colours: black, brown or even red. Their growth is largely superficial, and so their presence is not in itself detrimental to the structural performance of the timber. However, their continued growth does suggest the distinct probability of high humidity and condensation. This can be of significance in the diagnosing of dampness problems and of condensation in particular.

Wood-boring insects

The presence of wet or dry rot decay is essential for some wood-boring insects, in particular the deathwatch beetle (*Xestobium rufovillosum*) and wood-boring weevils. Although this would appear to make the infestation of timber by these insects of secondary importance, they do increase the damage caused by extending the effects of even slight decay. The decay process itself will soften the wood and improve its nutritive status by, to some extent, pre-digesting it, and the presence of the fungal hyphae in the wood will increase the nitrogen content, an essential element for insect protein, which is otherwise only at a low concentration in many timbers. For this reason, many types of insect infestation develop more rapidly in partly decayed wood.

All wood-boring insects in the UK have a typical four-stage life-cycle in which the female beetles lay eggs on or in the timber. These hatch and the young larvae bore into the wood. The tunnels they make are filled with bore dust or frass, often containing faecal pellets, which have a characteristic shape and size that is useful in identification. The shape and size of the tunnel that larvae make in the wood is also characteristic in many cases. The full-grown larva pupates, usually near the surface of the timber, and emerges as an adult insect through a flight (exit) hole on the surface of the wood during the summer months. After mating, the adult female insects lay their eggs on suitable timber to continue the infestation. Most adult wood-boring insects can fly, and this obviously assists in the spread of the infestation. Apart from the boring of the 'exit' hole, the adults of most species cause little or no damage to the timber, and their role is restricted to reproduction and distribution.

Even for species that do not require the timber to be decayed in order to attack and complete their life-cycles, the wood moisture content can still be very significant. Certain insects can only attack green timber down to 30% moisture content of fibre saturation point. Conversely, under unfavourably low moisture levels the life-cycles may be considerably prolonged, possibly because of the extra hardness and low nutritive status of such dry wood. Longhorn beetle larvae are well known as a group to survive low-moisture conditions, and there are well-documented records of forest longhorn larvae emerging from furniture timber after many years of use.

Similarly, the common furniture beetle *Anobium punctatum*, the most widespread wood-boring insect in building timber, is more commonly found in situations where higher-than-average moisture contents might be anticipated. Examples of such situations are: timber in an upstairs cupboard; in kitchen, bathroom and WC floors; and timber moistened by condensation water under a roof cold water tank.

Most of the common species of wood-boring insects can be identified by examining the area and the species of wood attacked and carefully examining the bore dust for faecal pellets (BRE Digest 307). Further details are found in Table 4.5 and in the BRE guide to identification (Bravery, 1991). In many cases it is not essential to identify the species of insect, because remedial treatment techniques are similar. However, an identification can confirm that the insect is a species that does not require treatment, such as *Ernobius mollis* and Ambrosia beetle.

Preventive measures

Background

Clearly, even from this brief description of the effects of moisture on wood and fungal and insect deterioration of timber, it seems desirable

Table 4.5 Identification of some common species of wood-destroying insect found on building timber.

Species of insect	Damage to wood	Visual examination of frass	Finger test by placing sample of frass in palm of hand and rubbing with finger	Using hand lens (× 10)
Common furniture beetle (Anobium punctatum)	Exit hole and tunnels 1–2 mm diameter	Granular appearance. Individual pellets just visible	Slightly gritty: like very fine sand	Lemon-shaped pellets pointed at one or both ends
Deathwatch beetle (Xestobium rufovillosum)	Exit hole and tunnels 2–3 mm diameter	Bun-shaped pellets, about 0.75 mm in diameter. Easily recognised individually	Gritty: like coarse sand	Bun-shaped pellets in powdered material
Bark beetle (Ernobius mollis)	Exit hole and tunnels 1–2 mm diameter	Bun-shaped pellets. Similar to that of deathwatch, but smaller (0.5 mm). Usually dark in colour, being composed wholly or mainly of bark, but may be white or mixed, when boring in wood	As for deathwatch beetle, but less gritty	Bun-shaped pellets. Smaller than deathwatch
Weevil (Pentarthrum huttonii Europhryum confine E. rufum)	Exit holes 1.5–2 mm diameter. Oval-ragged or indistinct. Irregular chambers	A fine dust with a very fine granular appearance	Slightly more gritty than Lyctus but still rather fine	Very small pellets. Just possible to make out shape as being round-cylindrical. No dust
House longhorn (Hylotrupes bajulus)	Exit hole and tunnels 4 × 7 to 7 × 11 mm	Fine dust mixed with barrel-shaped pellets about 1 mm	When well rubbed, the pellets stand out and feel coarse	Large cylindrical pellets, with well squared ends in fine dust
Powder post beetle (Lyctus sp.)	Tunnels and exit holes 1–2 mm diameter	Very fine dust. No recognisable characteristics	Like rubbing talcum powder	Very fine powder. No characteristic shape. Frass composed entirely of such dust
Ambrosia beetles (Pinhole borers)	Exit holes and tunnels 1–2 mm diameter	No frass. Tunnel walls and adjacent wood usually stained black	N/A	N/A

that, as far as possible, timber be protected from all forms of dampness (Eaton & Hale, 1993). In particular, timber needs to be protected from penetrating dampness and the worse effects of condensation (Oliver, 1985).

There are several procedures that may be adopted to achieve these objectives, especially the use of physical moisture barriers and surface coatings (BRE Digest 304). The alternative is to use timber less susceptible to decay fungi, either because it is naturally durable or because it has been treated with preservative (BRE Digests 327 and 370). If no precautionary measures are taken, remedial treatment, often involving replacement of decayed material, may be necessary.

Natural durability

Naturally durable timbers are so described because during the growth of the tree they acquire a natural resistance to decay by virtue of the presence of chemical deposits laid down in the wood, the so-called 'extractives'. This may occur on heartwood formation, but the extent of natural durability varies both between different species and between different timbers of the same species, and even in different parts of the heartwood. A classification of natural durability in terms of the life expectancy of a 50 × 50 mm timber in ground contact is well established, and a durability classification has been developed for the more commonly used building timbers (Table 4.6). It is important to appreciate that no sapwood is durable, and consequently, if durable timbers are selected and a good performance in the decay situation required, then sapwood must be excluded or preservative treatment used. The use of durable timbers has been accepted in the Building Regulations.

Moderately durable timber or better is permitted for external weatherboarding on buildings. Durable timber is often selected for sills, door thresholds, weatherboarding and external joinery, particularly on buildings where a high standard of finish is required (BRE Digest 296). The use of hardwood timber in replacement double-glazed joinery is a topical example of the use of durable hardwoods to reduce decay risks.

Wood preservatives: pre-treatment

An alternative if not additional measure is the employment of wood preservative treatments, either before installation or as remedial in-service application.

There are two main traditional types of preservative used for the pre-

Table 4.6 Classification system for natural durability of timber in ground contact.

	Perishable	Non-durable	Moderately durable	Durable	Very durable
Life expectancy of 50 mm stake in ground contact (years)	0–5	5–10	10–15	15–25	25+
	All sapwood	Firs (Abies)	Douglas fir	Western red cedar	Afrormosia
	Abura	Western hemlock	Larch	Agba	Afzelia
	Alder	Baltic redwood	Maritime pine	Sweet chestnut	Ekki
	Ash	Whitewood	African walnut	American mahogany	Greenheart
	Balsa	Spruce	Keruing	Meranti	Iroko
	Beech	Parana pine	African mahogany	Oak	Jarrah
	Birch	Elm	Sapele	Utile	Kapur
	Poplars	Obeche	Seraya	Idigbo	Lignum vitae
	Willow	Podo		Karri	Makore
				Pitch pine	Opepe
					Teak
					Guarea
					Mansonia

treatment of building timber: the waterborne copper chrome arsenic mixtures (CCAs), pressure-applied to BS 4072, and the organic solvent double vacuum treatments to BS 5707. Use-orientated specifications are also published that enable different preservative systems to be selected for particular end-uses in buildings (BS 5268 Part 5 and BS 5589).

For the timber to be satisfactorily treated with a wood preservative, certain preconditions should apply. It should be:

(1) at the correct moisture content (below 25% preferably);
(2) free from pests, decay and insect attack;
(3) free from bark, paint, dirt or glue – in fact anything that might interfere with the preservative penetration;
(4) in its final shape and size – if pre-treated timber is cut or machined after treatment, the exposed surfaces should be retreated;
(5) a relatively permeable species of timber.

The permeability classification of timber has been developed by measuring the average lateral penetration of preservatives into timber after two to three hours' pressure treatment. Generally, it is desirable to select timber from the permeable or moderately resistant categories (Table 4.7), because the better penetration obtained should provide a better performance in service, especially in the more severe environments.

The waterborne preservatives are based on CCA salts, and are usually

Table 4.7 Classification of the permeability of common timbers.

	Permeable	Moderately resistant	Resistant	Very resistant
Lateral penetration (average), 2 to 3 hours' pressure treatment	Complete	6–18 mm ($\frac{1}{4}$–$\frac{3}{4}$ in)	3–6 mm ($\frac{1}{8}$–$\frac{1}{4}$ in)	No lateral penetration
	Sapwood Alder Beech Birch Sycamore Podo Ramin	Abura ash Opepe Pine Elm	Balsa Keruing Western hemlock Obeche Spruce Larch Western red cedar Whitewood Douglas fir Fir	Sweet chestnut Ekki Kapur Oak Greenheart Teak Jarrah Karri

applied by pressure treatment methods (BRE Digest 378). The normal full cell cycle (Fig. 4.8) consists of an initial vacuum stage during which some of the air in the wood cells is removed. The preservative is then run into a tank and pressure is applied up to 10 bar (1 bar = 10^5 N/mm^2) or 150–200 lb/in^2 for a sufficient period until a gross retention is obtained. When this is achieved, pressure is reduced and the preservative is removed from the treatment cylinder prior to a final vacuum period, when excess preservative is extracted from the wood and a drier timber is produced. Even so, the timber should always be allowed to dry for several days before use. During this time, the moisture content should be reducing towards the level it will achieve in the final use, and fixation of the chemical salts in the wood can take place. This post-treatment drying is desirable if movement of the timber in use, caused by subsequent drying, can be minimised.

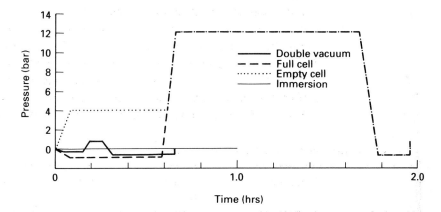

Fig. 4.8   Pressure treatment and double vacuum treatment cycles for timber preservation (after BRE Digest 378).

For these reasons, CCA preservative treatments are generally preferred for structural timber and not for joinery, where the redrying and movement would make the assembly of machined joinery components difficult. In the manufacture of truss rafters, where galvanised metal plates are used to form the joints, there can be a possibility of corrosion, either when trusses are made from freshly CCA-treated wood that has not been completely dried, or if the trusses are re-wetted on the building site. For these reasons, probably the major proportion of building timber that is pre-treated today is processed using organic solvent preservatives (Fig. 4.8). With organic solvent preservatives, fungicidal and insecticidal active ingredients are dissolved in a white spirit type solvent and applied by one of several cycles (BS

5707). The most usual fungicides used are either 5% pentachlorphenol or 1% tri-butyl tin oxide (Fig. 4.8). Lindane (0.5% hexa-chlorocyclohexane y HCH) was a very common insecticide, but is now no longer used in remedial works because of its known noxious properties. Its use in pre-treatment is also declining.

While organic solvent preservatives are initially more expensive than the waterborne types, they have several practical advantages. The dry condition of freshly treated timber means that the processing can be incorporated more easily into timber-manufacturing processes. This is because the organic solvent preservative does not cause any alteration in the moisture content, or the timber dimension in machined components, and the timber can be glued or painted soon after treatment.

Borates are used as a fire retardant as well as a preservative for timbers, principally as a fungicide. They can also be used as an insecticide, but not for on-contact purposes.

The specification of pre-treated building timbers is covered by BS 5268.

Water-repellent treatments

Organic solvent wood preservatives can be formulated with water-repellent additives, which can reduce the rate of water uptake by treated wood. Various water repellents are used, including alkyd resins, silicones and waxes. These treatments impart a water repellency to surfaces by coating the cell walls in the treated zone. The penetration of water is retarded because it is repelled from coated surfaces, and water penetration by capillarity is also reduced. They have no effect on water vapour movement, so the drying of wet wood is not affected.

These treatments are particularly effective in improving the long-term performance of surface coatings by reducing the extent and speed of movement of the underlying wood substrate. Thus if water does penetrate through the surface coating, it is repelled and does not penetrate as rapidly.

Water-repellent treatments to timber prior to painting or other surface coating application have not been widely used, possibly because the extra cost has deterred buyers.

Research by the BRE has indicated that the performance of some water-repellent treatments may decline over time (BRE Digest 354). Investigations are still proceeding to find out why this is so, and whether there are any particular formulations that are more durable.

Weathering of timber

When timber is used on the outside of a building – as joinery, barge-boards, fascias, soffits, or as weatherboarding, for example – changes occur to surfaces over the months and years of weather exposure. Moisture in the form of rain, snow, frost and water vapour, and sunlight, especially ultraviolet light, and mould, all combine to cause these changes, which are usually described as weathering. The most obvious result of this weather exposure is the loss of the original colour of the timber, accompanied by a gradual darkening and slow deterioration of the wood surface. Weathered timber is usually a silver-grey colour.

Any erosion of the wood surface is very slow, and a loss of about 6 mm from the surface as a result of weathering over 100 years has been quoted in the USA (see BRE Digest 296).

Normally, exposed timber on the outside of buildings will have an equilibrium moisture content level in the range of 20–25%. This is usually too low for decay in such wood to develop. Decay in external joinery therefore is much more likely to occur when the moisture content rises above 30%. This can happen if the design of the timber component or the use of coatings or other impermeable barriers restricts the drying out of wet timber. In some situations coating may actually cause persistently high moisture contents to develop, and decay follows.

Surface coatings

General

Paints, varnishes and stains are formulated to protect timber from these weathering effects. They are also used to enhance the timber's appearance by preventing the development of the weathered surface, or by masking an already exposed surface with a more attractive colour.

There are several different coating systems that have been developed for use on external timbers. These included the more traditional oleoresinous three-coat (gloss) paint system, external acrylic-type emulsion paint, the exterior pigmented stain and the newer 'microporous' paints. Each system is designed to cope with the combined problems of moisture penetration and movement, and to give weather resistance at a low maintenance cost (Table 4.8).

Table 4.8 Some active ingredients used in wood preservatives.

Biocide	Normal concentration	Application (pre-treatment)	Concentration in remedial products	
			Organic solvent	Emulsion
Copper Chrome Arsenic	2–5%	Pressure	N/A	N/A
Fungicides				
Pentachlorphenol	5% m/m	Immersion and double vacuum	5% m/m	5% m/m
Tri-butyl tin oxide	1% m/m	Immersion and double vacuum	1% m/m	1% m/m
Trihexalene glycol biborate (Borester 7)			1.0% boric acid equivalent	
Insecticides				
Dieldrin	0.5% m/m	Immersion and double vacuum		
Lindane (γ HCH)	0.5% m/m	Immersion and double vacuum	0.5% m/m	0.5% m/m
(γ Hexachlorocyclohexane)				
Permethrin	0.1% m/m	Immersion and double vacuum	0.2% m/m	0.2% m/m

Three-coat oleoresinous paint system

This is the traditional and long-established method of finishing external joinery and other timber, and usually consists of primer, undercoat and top coat. It results in the formation of a film on the surface of the wood up to about 0.1 mm in thickness and can significantly influence water uptake. It is particularly effective in this respect if applied to end grain areas and the backs of components prior to fixing. As well as retarding water penetration, oleoresinous paints provide a barrier to water vapour, and can offer the potential to retard moisture content fluctuations in the painted wood to a significant extent (BRE Digest 354).

The primer is designed to penetrate into the wood, to provide a base for subsequent coats and to provide a degree of protection from moisture pickup during storage on site. Primers with aluminium leaf pigment (to BS 4756) are probably the most effective moisture sealants.

Undercoats have a higher content of resin and oil for 'film build' and a higher pigment content for opacity. Top gloss coats have more medium (oil or resin content), which dries to form a weather-resistant glossy top surface.

Oleoresinous paints are based on a medium, usually combining linseed oil and alkyd resin, thinned with white spirit. The paints also contain finely ground inorganic pigments that provide the colour, opacity and protection from ultraviolet radiation. The traditional lead-based pigments have been replaced in recent years by titanium and zinc-based pigments, which are regarded as 'less toxic' than lead. This change seems to have resulted in the greater incidence of staining fungi on paint and varnish films on wood.

Several species of staining fungus have been recorded growing on and in external painted woodwork. *Pullularia pullulans* is the most commonly quoted species. The term 'blue stain in service' has been coined to describe this effect. The staining fungus causes a blue discoloration of the wood and hyphal growth on the surface coating itself. A breakdown of some of the oil resins present in the coating occurs, and the fungus growing on the wood develops spore pustules, which rupture the surface coating and allow water penetration to occur. The wood substrate becomes preferentially wetted and the coating soon becomes detached.

In the more recently formulated paints, the importance of the growth of these staining fungi has been appreciated, and a new range of fungicides are incorporated in the coating system.

A particular problem arises when 'weathered' wood is recoated. The fungus stain is already established in the wood, and it is important to try to prevent it from recolonising the wood surface and the new coating film. 'Preservative' base coats have been developed to control the staining fungus already in the wood.

The actual weathering process of unfinished timber also causes deterioration of the surface layers of the wood cell structure as a result of degradation by ultraviolet light. This makes the surface weaker, and increases the difficulty of obtaining a sound 'key' on the wood substrate. This emphasises the importance of regular maintenance of external painted timber before staining mould develops and ultraviolet deterioration can take place. Repainting of exterior coatings using oleoresinous materials normally should take place every three to five years, depending on the exposure.

For an external paint film to provide a degree of control of timber moisture content fluctuation, correct application is essential. For the factory application of coating systems to machined joinery components, it is important that the timber is machined to the final shape and size, and all surfaces are accessible for priming prior to assembly. Furthermore, the timber should be at an appropriate moisture content, close to that of the equilibrium moisture level that it will achieve in service.

As has been described earlier, the uptake of water by the end grain of timber can be considerable and rapid. In many external joinery uses of timber, the end-grain application of primer is overlooked, and penetration of water can swiftly ensue. The situation is then compounded by the presence of a paint on the side-grain surfaces, which reduces the rate of drying that might otherwise occur through these exposed surfaces. The necessity of back priming and coating of timber in contact with potentially or actually wet masonry is also often overlooked. This aspect is further discussed in Chapter 5. The net result is high moisture content in external joinery timber, and the onset of wet rot decay.

Even before decay develops, fluctuating wood moisture content and wood movement put great stress on the surface coatings. When the sun shines, the surface paint heats up and expands, but the underlying wood may dry a little, causing lateral shrinkage. When rain follows, the wood may expand but the surface paint may contract because it has now cooled. The surface coating is formulated to be flexible and to accommodate these stresses, but in time flexibility is reduced and checking and cracking may follow. Ultimately the coating may become detached by flaking or peeling from the wood surface when adhesion is lost.

Desirably, maintenance recoating, through a planned preventive scheme, should precede the detachment of the surface film. But if it does not, then sanding and burning-off of the remains of the paint down to the base wood is usually necessary before recoating can begin.

Unfortunately, maintenance recoating of painted timber is frequently delayed too long. Thus because the costs of stripping and recoating are high, alternative strategies and different types of paint and stain coatings have been developed in recent years.

Emulsion paints

An emulsion paint consists of a pigment and a resin binder, emulsified and suspended or dispersed in a water carrier. These paints have been widely used for many years on interior plastered wall surfaces. The binder used for many years is usually polyvinyl acetate (PVC), but when copolyermised with an acrylic resin, emulsion paints can perform well on external surfaces. BS 5802 provides a specification for such products for use on exterior timber.

Acrylic primers and undercoats are quick drying, and this facilitates rapid application in the joinery factory. However, it is not yet possible to produce a full 'gloss' paint of this type, and 'eggshell' or 'semi-matt' is the best that can be achieved. Consequently, emulsion paints are sometimes overcoated with a gloss oleoresinous top coat to provide better weather resistance.

A paint coating of this type will be more permeable to moisture and water vapour, as emulsion coat particles 'coalesce' but do not form such a continuous barrier to water vapour as the oleoresinous paint film does. If wood becomes wet, through either water or vapour penetration, quicker and greater drying can be expected during drier weather periods, reducing the time during which the external timber is at risk from decay.

Clear and translucent finishes

The appearance of wood is one of the most attractive and natural finishes. This can be enhanced and restored by the use of clear finishes (BRE Digest 387).

Some years ago, there was a vogue for using timber externally as cladding on buildings and coating it with a clear varnish. Varnishes at that time were formulated in a similar way to the oleoresinous paint, but without any pigment, of course. It was soon found that varnishes performed very poorly on external exposure, possibly offering only half the 'life' that was expected from paint. A fashion for polyurethane-based varnishes was equally short-lived, because these were even less flexible than the alkyd or phenolic resins normally used in varnishes.

A partial 'breakthrough' was achieved with the incorporation of siliceous ultraviolet absorbers, first developed at the BRE as 'long-life varnish'. When a mouldicide was incorporated, it achieved the best possible performance for a clear varnish on timber. However, without the advantage of a pigment to resist the effects of both visible and ultraviolet light, varnishes offer a relatively short service life (Table 4.9), which necessitates frequent maintenance and recoating, and extensive stripping and burning-off of the original coat is necessary before recoating can be carried out.

In the search for a coating that allowed the natural wood appearance to be seen yet had simpler recoating and maintenance procedures, the concept of an exterior wood stain was developed. These products were designed to penetrate further into the wood surface, enhancing the appearance with a range of attractive pigments held with a binder of water-repellent or resin in an organic solvent such as white spirit. The overall formulation was of a lower solids (pigment and binder) content. A fungicide was incorporated to provide a degree of control of all mould growth. Water was shed from treated wood surfaces by the water-repellent properties obtained from silicones, waxes or resins and oils incorporated as water-repellent and binders.

As for emulsion paint, the concept was to allow the wood to 'breathe'. Wet timber would dry because the surface coating offered less control of water vapour movement. Because the coating thickness was so modest, failure was a gradual affair of colour loss without flaking, cracking or peeling of the surface coating.

Recoating was also a simple matter because a reasonable appearance can be achieved even over the roughest weathered surface, which, if dry, would absorb much more of the finished solution. There is some evidence that such coatings actually perform better on 'rough-sawn' surfaces than on the planed surface for external timber joinery, probably because of the greater surface area and thus better potential absorbency of the former over the latter.

Microporous paints

The latest candidates for consideration as an external coating from external joinery are the 'microporous paints'. Various manufacturers, by different routes, are attempting to combine the best features of stains, emulsions and oleoresinous paints.

Some microporous paints are stains with higher solids contents, being organic solvent based; others are water based and emulsion-like in character. It is still too early to determine which type is going to perform well on external timber. Recent tests on microporous paints suggest that

Table 4.9 Some properties of external coatings for timber.

	Oleoresinous paint	Oleoresinous varnish	Emulsion paints and stains[a]	Water-repellent stains[a]
Solvent	White spirit type	White spirit type	Water	White spirit type
Application	3 coats	3/4 coats	3 coats	2/3 coats
Finish	Glossy	Glossy	Matt or semi-matt	Matt to semi-gloss
Moisture barrier	Good	Average	Low	Low but water-repellent
Film build	High	High	Medium	Low–medium
Life expectancy	5 years	2–3 years	2–3 years	2–4 years
Maintenance period and facility	Regular or difficult	Frequent or difficult	Frequent and easy	Regular and easy
Fungicidal	Possible	Possible	Possible	Yes
Elasticity	Lost with time	Lost with time	Good	Good
Types of failure	Erosion Checking Detachment	Checking and detachment	Erosion	Erosion

[a] Products of these types may be described as 'microporous'.

some may indeed not be very much more 'porous' to water vapour than more traditional coatings (BRE Good Building Guide 22).

Permeable paints will permit high moisture contents to develop for long periods. The current emphasis is towards the importance of end sealing and back priming components before fixing, and the use of coatings with more elasticity that can cope better with wood movement and the stresses associated with the opening of joints.

A new stain, organic solvent based, has recently been launched, which claims to combine the advantages of both acrylic and alkyd resin systems. An indication of the performance of the main types of coating system used on exterior timber is given in Table 4.9.

Of course, these new developments do not assist in dealing with timber components already in place where ends or backs are inaccessible. In such situations some form of in-situ injection of preservative with water-repellent might prove to be a satisfactory remedy (see Chapter 9).

Chapter 5
Dampness and Main Building Elements

Building morphology

Dampness in the main elements of a building's structure can arise from a variety of sources, but the principal causes being condensation, rising damp and penetrating damp, and these will be looked at in the following chapters. The major topic of this chapter, however, is the performance of the external fabric, roofs, walls, and floors in resisting or tolerating dampness. It will not only deal with traditional buildings but will also consider dampness problems in modern construction.

In basic terms, building forms can be categorised into two main groups:

❏ Traditional construction:
- solid load-bearing masonry with pitched roofs;
- load-bearing cavity brickwork with either pitched or flat roofs;
- cruck or balloon frame timber construction with brick or wattle and daub infill panels, and pitched roofs.

❏ Modern and non-traditional construction:
- skeletal framed construction in concrete, steel with concrete, with cladding or infill panels and pitched or flat roofs;
- steel or concrete portal frame construction with corrugated or profiled cladding;
- timber platform framed construction with brick or block outer skin, and pitched roofs;
- RC large panel construction with flat roofs;
- modular construction with overcladding and flat or pitched roofs.

Rainwater penetration and rising damp are the more likely common causes of moisture problems in traditional buildings. However, it has been pointed out by Howell (1995) that the incidence of the rising damp is not as extensive as commonly believed not only by the public but by surveyors as well (see Chapter 7).

In contrast, condensation and rainwater penetration are the primary forms of dampness problems in modern buildings. Why this is so depends on (a) the form of construction; (b) the materials used; (c) the microclimate; and (d) living patterns.

Most buildings are not made under factory conditions, and therefore contain a variety of inherent weaknesses as far as the precision of jointing, fixing and other weathering precautions are concerned. This is the main reason why it is not possible to prevent moisture from entering a building entirely. Frequently, attempts to block the transfer of moisture through the fabric of a building using impermeable materials are unsuccessful. Moreover, they may be counter-productive, because they can prevent moisture being dissipated, resulting in high moisture levels and decay to adjacent materials (Hutton *et al.* 1991).

Roofs

General

The function of a roof includes the provision of a durable, weatherproof and insulated cover for the building. Two main categories of roof can be differentiated: pitched roofs and flat roofs. Flat roofs are defined as having a slope of not more than 10° (BS 6229). Pitched roofs therefore are those having a slope between 10 and 75°. A pitch greater than 75° can be considered more as a wall surface than as a roof slope.

A roof is the most exposed part of a building, and thus is highly susceptible to the adverse effects of the weather. It has to resist not only precipitation but also snow, solar radiation, storms, and even user abuse. All these mechanisms can lead to deterioration of the coverings and eventually to leaks. Consequently, great care is needed in the design of new roofs as well as in the upgrading of existing ones.

As well as preventing penetration of rain and snow, provision must be made for shedding water clear of the building through gutters and downpipes. Generally, external drainage performs better and gives less trouble than any internal arrangement.

All projections through the roof can be regarded as sources of potential weakness. Thus special care is required in the design and fixing of flashings, dpc's and valley gutters to avoid rainwater penetration. This is particularly so where bunching of projections such as vent pipes through roofs takes place. It can be very difficult to provide adequate flashings between such pipework above the roof covering. Consideration should be given to either spreading such projections over the roof surface or combining the outlets into a larger vent pipe.

Pitched roofs

Pitched roofs have coverings that are usually watershedding in nature. In other words, a totally watertight seal is not necessarily intended. Rather, the aim is to shed rainwater and snow to a convenient discharge system.

Pitched roofs with overhanging or projecting eaves have much better weathering characteristics for the wallhead than do flush or clipped eaves (Fig. 5.1). In addition, pitched roofs with projecting eaves tend to be more aesthetically pleasing and allow for perimeter ventilation. Clipped eaves, however, may be appropriate for buildings in exposed areas susceptible to strong winds.

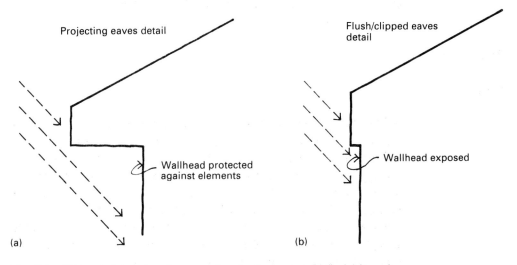

Projecting eaves detail

Flush/clipped eaves detail

Wallhead protected against elements

Wallhead exposed

(a)

(b)

Fig. 5.1   Different styles of roof eaves: (a) projecting eaves; (b) flush/clipped eaves.

Traditional pitched roof coverings are tiles, slates, stone, shingles or thatch, as the watershedding coverings. Some of these coverings are impervious, such as slate and tile, but their joints are susceptible to water penetration as a result of capillary attraction. Others, such as thatch, absorb some rainwater. Given that most of the water on a roof slope inevitably discharges to the eaves, it is the coverings at this part of a pitched roof that are most susceptible to moisture-related damage such as frost attack and delamination.

The lapping of joints in tiles or slated roofs must be sufficient to prevent rainwater penetration during periods of strong winds. Generally the flatter the pitch of a roof, the greater the lap that is required and the longer the drying out period. Slates and plain tiles require a double overlap to shed water from a roof. Single-lap tiles must be interlocking. Plain tiles should be laid on a pitch not less than 40°, and a 65 mm lap is considered sufficient except on severely exposed sites, where 75 mm is required.

An increased lap is required because water can be driven 'uphill' by the wind into joints and over upstands. A steady 60 mph (100 km/h) wind can support a column of water above 50 mm, and a roof must have the ability to withstand this. Also, snow can be driven under tiles, particularly at abutments with 'watergates' or secret gutters, and this can be a major problem in unlined roofs without any sarking. Any sarking must discharge clear of the external walls.

The fixings of tiles and slates, too, must be carefully considered. Modern artificial slates, for example, are thinner and lighter than natural slates. As a result, if artificial slates have only one fixing instead of two per unit, the coverings will be susceptible to 'chattering' or even loosening.

The use of dry-fix (rather than mortar-bedded) cappings to ridges and verges can minimise damage to the roof coverings during storms. Mortar as a 'wet-fix' method can accommodate slight unevenness in the roof covering, but is susceptible to cracking, and can become loose over time.

Lined roofs, especially those insulated at ceiling-joist level (i.e. 'cold' pitched roofs), are prone to condensation in the loft area and frost damage to bare pipework unless they are correctly ventilated and any water pipes and tanks suitably lagged (BRE Digest 270). In addition, efflorescence can appear on the timbers in such roofs' spaces if they are unventilated. Building regulations require cross-ventilation by means of permanent vents at eaves level, equivalent in area to a continuous gap along two opposite sides of 25 mm in width, plus a continuous gap along the ridge of 5 mm in width.

Where mansard-type roofs are constructed, or where sloping ceilings and walls consist of plasterboard attached directly to the lower or inner surface of the rafters, there is a vital need for ventilation (50 mm minimum gap), especially if insulation is laid behind the plasterboard. In effect the loft space becomes a 'warm' pitched roof, but the rafter profile is equivalent to a 'cold-deck' roof.

Attack by frost is more likely on tiles and slates on flatter-pitched roofs (i.e. between 25 and 40°), and the coverings take longer to dry out. Frost damage is more common at the eaves and other areas of the roof such as abutments where snow and ice may accumulate. The risk of frost damage is further increased where roof insulation has been installed. An easy way of determining which roofs have been insulated is to see whether or not snow is still on the roof the day after a heavy snowfall. If it has quickly melted off, it is likely that the roof is uninsulated, and the heating inside the house has escaped through the roof unchecked, melting the snow on its way. Thermographic aerial surveys, however, are much more accurate at detecting such heat losses (Hollis, 1991).

Shingles

Shingles are tile- or slate-like coverings for roofs and walls. In the UK they are usually of riven or sawn thin sections of wood (Wright, 1991). In other countries, however, such as the USA, felt and asbestos cement shingles are commonly used. Wooden shingles in the main were originally of oak, but since the 1930s Western red cedar has been used in the UK. Shingles as a roof covering and wall cladding are more common in the countryside, and are found in relatively small dwellings or agricultural buildings. Such roofs should have a minimum pitch of 30°. In rural areas and north-facing surfaces moss and lichen can grow on shingles, and these organic growths maintain continuously wet conditions that even double cedar cannot withstand. Mild steel and even galvanised fixings are susceptible to corrosion in shingled covered roofs because of either acidic preservatives or resin exuded by the timber. These problems can be minimised by using an alternative to organic salt preservative treated shingles (such as a borate preservative), fixed with copper or stainless steel nails, on roofs designed with steeper pitches.

Thatch

Thatch forms a roof covering, made of dead plant material other than wood, which provides good thermal and sound insulation (Hall, 1988). It is, like shingles, usually encountered only in low-rise rural buildings such as cottages, and barns.

Thatched roofs normally have a deep eaves overhang, which obviates the need for gutters as rainwater is discharged well away from the building (Fearn, 1985). However, birds, squirrels, rats and mice may cause damage once entry via the thatch is affected. Netting is fitted over the roof surface to prevent this. A thatched roof can be expected to last from 20 to 40 years, depending on the material used and on the quality of the workmanship (NBA, 1985). Reed thatch is believed to be the most durable (Tables 5.1 and 5.2).

Fire from external sources is the main hazard with thatch, and special precautions should be taken if it is likely to be affected by burning embers from solid-fuel fires. As a measure for reducing the fire risk, the reeds should be treated with a fire-retardant preservative. The use of an aluminium foil vapour check under the thatch to resist fire spread and

Table 5.1 Thatch types and performance.

	Norfolk reed (*Phragmites communis*)	Combined wheat reed	Long straw
Life (years)[a]	60+	30–40	25
Distinguishing features			
Appearance (roof shape profile)	Angular	Soft and rounded	Soft and rounded
Mechanical restraint	Not common	Wire or nylon mesh	Wire or nylon mesh
Spars	Not commonly visible	Not visible	Often visible at eaves edge
'Stalk ends'	Butt ends of reeds visible; irregular shape	Butt ends visible; regular ovals	2.5–5 cm stalks visible

[a] Life expectancy best on roofs with more than 45° pitch, away from overhanging trees, and in regions of lower rainfall (lower water penetration); ridge has to be renewed every 12–15 years.

Table 5.2 Stages in the weathering of thatched roofs.

Period (years)	Feature
New	Yellow colour
3	Colour lost on exposed surfaces; gullies become darker first
5–8	Dark bands develop under eaves close to the wall; these become thicker and irregular in time Dark triangles under the eaves indicate water penetration Wire mesh rusts
9	Slippages of spars at the ridge and at the eaves with long straw thatch
10	Wire mesh splits at the edge of the eaves
15	Brown/black colour

resist rainwater penetration during the inevitably long installation period has also been advocated. The danger with using such fire/vapour checks is that they restrict air movement in the thatch and do not allow it to dry out properly. This can be exacerbated where thatched roofs are situated near overhanging hedges or other trees, which can reduce the drying-out time because they restrict radiant sunshine on the roof.

If a thatched roof is not allowed to dry out properly, the outer layer will be highly susceptible to early deterioration, and fungal decay can result in the reeds. Thus it is important that thatched roofs are allowed to breathe properly. They are meant to act as a form of 'overcoat' weatherproofing.

The problems associated with thatched roofing are of such concern to the industry that it has commissioned the BRE's Scottish Laboratory to conduct a detailed study of the environmental conditions within and the performance of typical thatched roofs. At the time of writing an interim report by the BRE is in the process of being compiled.

Fixings

Whatever roof covering is selected – clay or concrete tiles, slate or shingles, for instance – it is important that their fixing nails have a service life comparable to that of the covering. Unprotected nails are generally at risk from rusting and corrosion. This form of failure is known as 'nail sickness'. Galvanised nails are not always totally satisfactory because there is a risk of damage to the galvanising during driving and fixing. Copper or stainless steel nails are more durable.

In a modern building, wood tiling battens and counterbattens are usually made of preservative-treated softwood, as they are probably at greater risk from decay when they are fitted above sarking felt, than in unlined roofs.

Sarking

In Scotland and the North of England the use of timber sarking boards on rafters onto which felt, counterbattens and battens, and then tiles are subsequently fixed is a common practice. It is also very long-established practice in Scotland to nail slates directly to the felt on sarking boards. The greater weight of many roof coverings used in the North and the greater need for insulation was no doubt responsible.

In England, sarking was not common on the average dwelling until required by building regulations in the mid-twentieth century, and sarking felt has been the material most often used. Such felt provides an additional watershedding layer beneath the roof covering if rain or snow penetration occurs, and it should be arranged so that any water that does enter is carried clear to the eaves.

The use of battens on felt on sarking board is very inadvisable because of the risk of wet rot to the timber battens if there is a roof leak. In such a case the water will pond behind the batten and eventually cause it to decay. That is why the use of counterbattens is essential in such situations.

Roof drainage

Clearly the gutters and downpipes should be adequate to accommodate the rainwater discharge from the roof. Calculations are usually based on a rainfall peak of 75 mm in an hour. In fact, statistics suggest that this occurs for a period of 5 min about every other year. Wind will affect the angle of rainfall, and from the design standpoint the angle is assessed as being 25°. The rate of run-off is obtained by calculating the roof area, and building regulations indicate the size of gutter and outlet that is appropriate for the water flow capacity predicted.

The flow capacity of a gutter depends on its cross-section, shape, length, presence of bends and overall gradient. Obviously, the positioning and number of outlets will affect the size of gutter required, and in the current Building Regulations the appropriate sizes are indicated.

There is always the risk that too great a gradient of gutter might cause it to miss some of the rainwater being discharged off the roof. When water leaves the edge of a slate roof, there is little spread, so the gutter can collect most of the water even when it is fixed some distance below the roof edge, though this should not in any event exceed 50 mm. Clay pantiles cause a wider spread of water from the edge. Where the lower edge is rounded, the water is deflected more to the rear (i.e. towards the wall). It is best for the upper surface to be rounded and the lower corner sharp.

Metal gutters may be affected by rusting and corrosion, either from atmospheric pollution or by 'contamination' of the water by the roof covering. Cast-iron gutters (or 'rhones' as they are sometimes called) and downpipes have been largely superseded by plastic ones in modern buildings. These are generally believed to have a life expectancy of over 20 years (HAPM, 1993), do not require any regular maintenance, and are cheaper to install.

One must, however, be careful about using the term 'maintenance-free material'. Strictly speaking, any material or product, no matter how high its quality, may be susceptible to damage or even unforeseen problems, and thus will require some 'maintenance' to keep it in a reasonable condition. This applies to plastics, stainless steel and other so-called 'maintenance-free' materials.

Correct installation and repair of any damage is essential if gutters and downpipes are to perform satisfactorily. Regular maintenance (e.g. removal of leaves and accumulated debris every six months) is vital. It is always a good idea to check the performance of gutters on buildings by carrying out an inspection during rainfall, when many defects can be seen more clearly. Particular attention should be paid to hidden gutters, such as those behind parapets and in roof valleys.

One of the most obvious symptoms of problems with roof drainage is either vegetation growths in gutters or algae/lichen staining on walls behind rainwater goods. Such symptoms indicate that the building has been poorly maintained and runs the risk of decay in timbers associated with soaking of the walls at or near such defects.

The adequate and correct disposal of rainwater below ground is described in the Building Regulations, where discharge capacities for rainwater drains of 75 mm diameter or greater over a range of gradients are quoted. Burst or defective below-ground drainage can be very damaging to a building. It could encourage subsidence to occur in the building by saturating (and thus softening) the ground around the foundations, and cause the water table to rise. This in turn could encourage rising damp to occur in a wall without a dpc, or in a wall with a defective dpc.

In existing buildings, soakaways are often encountered to discharge surface water, particularly in rural areas. The performance of these reservoirs will be affected by the characteristics of the subsoil and the height of the water table. It is important that surface water is carried an adequate distance from the building (at least 15 m), down the slope. The effectiveness of soakaways may be reduced with time as they become silted up.

Chimney stacks

The materials used in the construction of chimneys have to withstand more severe weather exposure than any other parts of the roof (Fig. 1.1), as well as the effects of heat of the fire and the properties of combustion of the fuel. At the top of the chimney, a good coping system should be incorporated to protect the masonry below. It should be waterproof, and should project 400 mm clear of the structure below to throw water clear. If the coping is jointed, a dpc should be set below it.

Moisture penetration may lead to salt and frost deterioration of mortar and bricks, and ultimately the moisture may flow down the stack to cause dampness and discoloration in adjacent materials in the interior as well as premature condensation of the flue gases, which are very acidic. Chimney masonry needs to be weather resistant, and damp-proof courses are essential. Indeed, all projections through the roof can be regarded as sources of potential weakness. Typical provisions for flashings and dpc's for chimneys are shown in Fig. 5.2.

Chimneys should be built with dpc's that prevent the downward passage of water to the interior of the building. The aim should be to

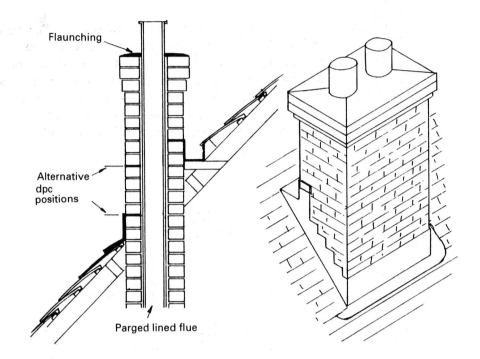

Flaunching

Alternative
dpc
positions

Parged lined flue

Fig. 5.2   Design of chimneys to prevent water penetration (flashings and dpc's).

provide a horizontal dpc through the thickness of the chimney with an upturn at the inner face, which, at the same time, is continuous with vertical flashings at the intersection with the roof. At the junction with pitched roofs, the dpc in the chimney should be stepped with, and be continuous with, the stepped flashing on the roof slope. It may also be necessary to install another dpc in the chimney stack in the roof space to prevent moisture penetrating into the masonry below the ceiling. Frequently, defects at this junction are given a quick 'remedial' treatment by applying a mortar fillet known as a cement skew, which is the common method of 'sealing' roof abutments at gables of old buildings. These fillets rarely perform in a satisfactory manner, as they are soon detached from the chimney and roof junction by thermal and moisture movements of the structure.

Many chimneys are rendered, probably in an attempt to protect the underlying masonry. If this render becomes ineffective or fractured, general deterioration would be expected, and might even occur faster than on unrendered chimneys.

Slow-combustion boilers burn for long periods at very low temperatures, and the heat generated in the flue is not sufficient to prevent condensation on the flue walls. The resultant deposit may contain sulphur compounds, tar, ammonia and soot, as well as water. This 'con-taminated' and highly acidic water is often absorbed by the rendered lining called pargetting. If allowed to penetrate through to the mortar and brickwork it will cause staining, and sulphate attack of these parts of the chimney can occur. This effect may be spread further by any pene-trating rainwater. The expansion of sulphate-attacked mortar causes chimneys to lean over, and may lead them to collapse (Fig. 5.3).

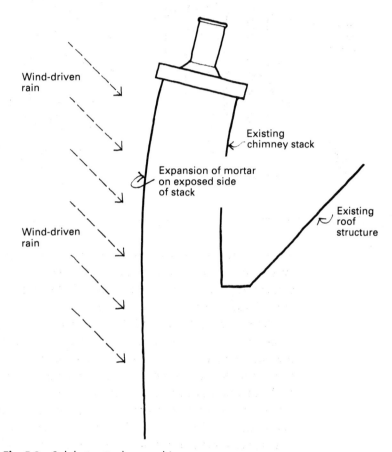

Fig. 5.3   Sulphate attack on a chimney.

The normal remedial action is to line the flues with an insulated impervious stainless steel liner, as well as to repair or rebuild the chim-ney, using a sulphate-resisting mortar and good-quality bricks. Of course, a liner will also be necessary to prevent flue gases from escaping through any defects in the chimney's masonry.

To avoid interstitial condensation and the worst effects of water penetration in redundant flues and chimney stacks, it is important that adequate ventilation is arranged at the chimney head and at the fireplace if it is blocked up.

Flat roofs

General

Flat roofs became increasingly popular in the period after World War II. The structure is usually either in timber, concrete, steel, or a combination of steel and timber or steel and concrete. The waterproofing is in the form of roofing felt, asphalt or sheet metal.

Although there have been many failures with flat roofs in terms of leaks and ponding, there have been considerable improvements in the performance since the early 1980s (BS 8217: 1994). Properly designed and constructed, flat roofs can perform as well as any pitched roof (CIRIA & BFRC, 1993). The former type of roof still has advantages over the latter as regards cost and its ability to accommodate large and complicated layouts.

Flat roofs should be designed to clear surface water as rapidly as possible, and a fall must be incorporated in the design (CIRIA & BFRC, 1993). A 1:40 fall is widely recommended for this, although many old flat roofs have inclines at the previous recommended ratio of 1:80 (CP 144). Provision must also be made to ensure that inaccuracies in construction do not allow deflections to occur that might result in the development of rainwater ponding (which is not uncommon on roofs having a fall of 1:80). The effect that any deflection of the deck might have on drainage must also be considered, as it is common to support internal rainwater downpipes near columns or walls and not at the lowest point of any possible deflection. Generally, outlets should be level with the roof surface, and should be situated at the lowest point/s on the roof. Square sumps 1 m × 1 m should be provided at outlets in large flat roof areas. Gutters should discharge rainwater effectively to the outlets.

The roof designer has to consider the level of insulation required and the risk of interstitial condensation occurring. Insulation in the form of organic boards (e.g. cork board), mineral fibre (e.g. glass fibre quilt), plastic foam (e.g. polystyrene board) or composite boards (e.g. corkboard/polyurethane) can be used (Euroroof, 1985). The three main forms of flat roof – cold roof, warm roof and inverted roof – are described in Fig. 5.4.

The rate of heat flow through a roof is determined by the thermal

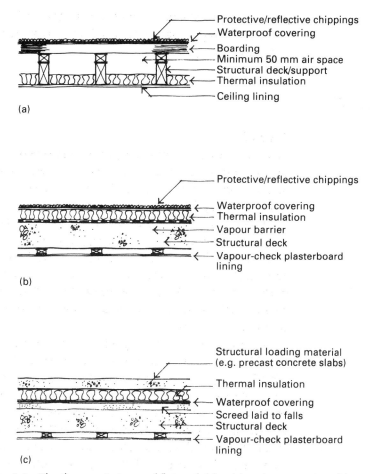

Protective/reflective chippings
Waterproof covering
Boarding
Minimum 50 mm air space
Structural deck/support
Thermal insulation
Ceiling lining

(a)

Protective/reflective chippings
Waterproof covering
Thermal insulation
Vapour barrier
Structural deck
Vapour-check plasterboard lining

(b)

Structural loading material (e.g. precast concrete slabs)
Thermal insulation
Waterproof covering
Screed laid to falls
Structural deck
Vapour-check plasterboard lining

(c)

Fig. 5.4   The three main types of flat roof: (a) cold roof construction; (b) warm roof construction; (c) inverted roof construction (protected membrane roof).

conductivity of the elements of the roof system, and the regulations set up mandatory U values for all main elements of a building (BRE Digest 324). These are values representing heat losses through various constructions, and in current building regulations in the UK a maximum U value of $0.35\,W/m^2K$ is required for the flat roof of a dwelling, but this is now dependent on the SAP rating of the property (see Chapter 6).

Cold-deck roofs

In cold roof design, insulation is positioned under the deck, and because heat is retained in the room below, the temperature level in the roof structure above the insulation is inevitably lowered (Fig. 5.4(a)). This will result in an increase in relative humidity in this area as well. In flat and low-pitched roofs, the volume of air above the insulation is small, and this increases the risks of saturation and condensation. For these reasons cold-deck flat roofs are no longer permitted in Scotland and are not recommended in the rest of the UK (NHBC, 1993, Chapter 7.1-D.1). They may only be used elsewhere in the UK where:

❑ the required level of ventilation can be achieved;
❑ ventilation paths are not blocked by structural or other members (e.g. solid strutting); and
❑ a ventilation space of 50 mm can be maintained.

To avoid decay problems that might arise as a result of moisture penetration into timber flat roofs, the National House-Building Council (NHBC) requires durable or pre-treated timbers to be used for joists and associated structural timber (Hamilton *et al.*, 1993; NHBC, 1993, Chapter 7.1-B/2; Powell-Smith & Billington, 1995). Decking must also be of durable materials such as pre-treated timber or plywood (to BS 6566). (Aluminium or galvanised mild steel profiled decking may be used, more usually on flat roofs of commercial buildings.)

Warm-deck roofs
In warm roof design, insulation is placed above the structural roof decking and beneath the waterproof roof finish (Fig. 5.4(b)). The vapour barrier is placed between the insulation and the deck (i.e. on the warm side of the construction). As there is always a possibility of condensation or leakage occurring at the vapour check, preservative-treated timber is recommended in this design (BRE Digest 312). There is usually no need to ventilate any voids formed in a warm roof.

Inverted roof
With the inverted roof system, insulation is placed above the roof waterproofing layers (Fig. 5.4(c)). The advantage of this system is that the waterproofing is not exposed to the excesses of temperature fluctuation, atmospheric pollution and foot traffic. As a result, expansion and contraction are reduced, as is the exposure to any aggressive agents. The insulation provided, however, has to be thicker to compensate for the wetting by the elements. The insulation also has to be protected from weather degradation and impact damage, usually using 'ballast' (i.e. rounded pebbles nominally between 30 and 40 mm or 50 mm thick paving slabs laid on special pads at the corners of each slab). In addition, before this protected layer is installed, it is usually prudent to carry out a water test on the roof membrane to locate and eliminate any leaks that may be present.

Precautionary measures for flat roofs

All flat roofs contain elements that expand, contract or move in relation to one another, and therefore subject the waterproofing to stresses. To avoid failure due to this movement, the waterproofing should ideally be isolated from the deck movement. This is achieved with prefabricated decking either by only partially bonding the first layer of waterproofing to the deck or by introducing an isolating insulation layer between the waterproofing layer and the deck (Briggs Amasco, 1983).

Blistering arises from pockets of entrapped air and moisture expanding under the heat of the sun. Pressures develop, causing displacement and stretching of the waterproofing, which may be irreversible. In general, blistering can be prevented or reduced in extent and frequency by the application of a layer of 12 mm white stone chippings, or two coats of aluminium solar reflective paint.

There has been a recent, understandable, tendency to reject the use of chippings because of the perceived risks of blocking drainage and the difficulty of tracing and repairing leaks covered with bonded chippings. It is of course essential to use gravel guards in all cases to protect rainwater drainage outlets that are less than 150 mm in diameter, and these will need regular maintenance (i.e. ideally, cleaning every six months).

Recent surveys suggest that modern flat roof coverings are more effective (BRE 372). Coverings such as high-performance polyester felts, single-ply polymeric membranes and metal sheeting, all of which have better flexural fatigue resistance, provide a relatively long, reasonably maintenance free life in the warm-roof configuration (CIRIA & BFRC, 1993). They also have the advantage of being self-finished thus obviating the need for chippings.

According to NBA (1985), the average life expectancies for the main flat roof coverings are as follows:

❑ traditional three-layer felt: 15–20 years;
❑ asphalt: 20–60 years;
❑ copper: 100–300 years;
❑ zinc: 20–40 + years (depending on pollution);
❑ single-ply uPVC/EPDM: 30–40 years;
❑ lead: 60–100 years.

Repairing flat roofs

Various procedures are available for repairing and re-roofing flat roof coverings. The selection of the method depends on the condition of the existing roof and the urgency of the repair (BRE Digest 351). There are a variety of quick-setting bituminous solutions that can be used to alleviate serious flat roof leaks. But this should be seen only as a stop-gap measure, not as a proper repair. Black bituminous coatings may improve a flat roof's water-resistance, but they increase its thermal absorbtivity (i.e. heat gain). Metallic-faced strips 100–150 mm wide coated with a bituminous adhesive can be very useful for carrying out temporary patch repairs to defective coverings or flashings. They are not suitable, however, for use with uPVC membranes because of the incompatibility between these materials, which could cause bitumen to attack the plastic.

If the existing waterproofing is in a sound condition and has no major blisters or ridges, then a single layer 'torch on' system using a high-performance felt with a mineral grit or metallic finish may be considered. Of course, it is necessary to remove all the stone chippings from the roof before the application of the new top surface. This can be very awkward and time-consuming.

It is, however, more likely that the opportunity to improve the roof's thermal as well as weathering performance will be taken. It is preferable to remove the existing covering back to the deck. If the deck is in a sound condition, or once any necessary repairs to it are completed, a new warm roof or inverted roof system can then be applied. If the existing construction was found to be of the cold-deck construction type, it would be advisable to take up the deck and remove any insulation between the joints in the case of a timber flat roof structure or remove the underslab insulation in the case of a concrete deck roof. The required falls of the upgraded roof can be achieved by using tapered polystyrene or cork boards on a new vapour check.

Walls

General

An external wall has to comply with many performance criteria: structural stability, weather resistance, sound and thermal insulation, fire resistance and durability, to name but a few. In the context of this chapter, however, the need to exclude moisture is of paramount interest. A more detailed analysis of moisture ingress resulting from rising damp and penetrating damp is given in Chapters 7 and 8 respectively.

Exclusion of moisture in walls is achieved either by employing materials that, while being permeable, are of sufficient thickness that penetrating water dries out before the inner surfaces are affected, or by providing in the construction a continuous cavity, which will interrupt capillary paths through which the moisture might travel (see Fig. 2.1).

Solid masonry walls (and thatched roofing) are examples of the first method of moisture exclusion. Cavity walling and rainscreen cladding are examples of the second type.

Solid walls

The walls of the majority of buildings constructed prior to World War I were of solid brick or stone construction. Such walls have generally performed well as far as weather resistance is concerned, especially if provided with generous overhangs and rendered external finish.

As has been discussed in Chapters 2 and 3, the various types of stone and brick vary in their resistance to moisture, but the general experience is that the performance of the mortar is probably more critical to the water resistance of the wall. The mix should always be selected to produce a mortar that is weaker than and as porous as, and certainly not stronger or less porous than, the masonry units. As mortar has an initial drying shrinkage and different thermal and moisture movements compared with many masonry units, the stresses that develop will sooner or later be relieved by the mortar cracking away from the units, and this effect is most marked with stronger mortars (BS 5224: 1993).

With stone walls, especially where impervious stone such as granite is used, penetration through the mortar may be more rapid because of the impervious nature of the stone. Monolithic concrete is also very dense and impermeable, and so penetration may occur only through the cracks, joints and local defects. Walls of concrete blocks and cast stone show greater thermal movement, and this can be dissipated better with weaker mortar mixes.

'No fines' is concrete mixed like ordinary concrete but without fine aggregate. Solid walls made from 'no fines' are usually between 250 and 300 mm thick, rendered externally with a two-coat render and finished internally with an insulation-backed composite plasterboard with integral vapour check (Fig. 5.5).

As the wall is full of spaces between the coarse aggregates, water penetration runs downwards on internal faces, and does not creep inwards, because there are no capillary paths. Provision must be made for this penetrating water to drain out of the wall at the base.

Attempts have been made from time to time to achieve some of the advantages of cavity walls by using brickwork laid to an unusual pattern such as 'rat-trap' bond (Fig. 5.6). With this method the bricks are laid on their side and a partial cavity is formed and linked by a header. However, this bond is more awkward to build and is usually more common in manhole construction than in buildings. If walls built in rat-trap bond

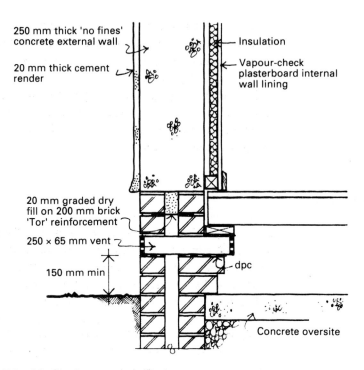

Fig. 5.5 'No fines' concrete walling.

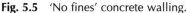
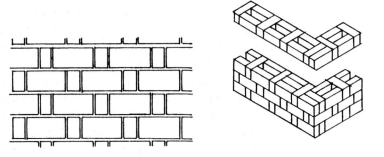

Fig. 5.6 'Rat-trap' bond.

are rendered, it may be difficult to detect such unusual masonry construction.

It may seem surprising that more penetrating dampness problems are not encountered in buildings built with solid walls. This is probably due in part to their less severe exposure compared with modern high-rise buildings, for instance. The use of more permeable lime-based mortars would allow a more general drying during dry weather periods.

Internally, lime-based plasters and several layers of wallpaper and paint may conceal some of the effects. Many of the external walls of such buildings would be dry-lined, and this would conceal some of the signs of penetrating dampness.

Current building regulations require that walls must resist the passage of rain or snow, and this requirement can be met by a solid wall of sufficient thickness covered externally with rendering or cladding, or by cavity wall construction.

Cavity walls

The cavity wall has proved to be the most effective design in the prevention of rainwater penetration, and the building regulations require that the cavity should have a minimum width of 50 mm. Experience has shown that the outer (cladding) leaf (Table 2.1) of the wall may sustain a higher degree of water penetration even under conditions of moderate rainfall, and a correctly constructed cavity wall will prevent this from penetrating the inner (loadbearing) leaf. With wind pressure, the extent of the water penetration can be considerably increased.

Most of the water penetration occurs through the joints between the mortar and the masonry units, and saturation will occur more quickly in walls with less permeable masonry units. The profile of the mortar joint is very significant in resisting water penetration. Bucket handle (keyed) and struck (weathered) profiles are the most effective in resisting penetration, and flush joints and recesses should be used externally only in relatively sheltered situations (Fig. 5.7).

Failures of cavity walls can usually be associated with bridging of the cavity, and this is frequently associated with the squeezing out of mortar from the joints that occurs during bricklaying. The best building practice ensures that such mortar is removed during bricklaying using timber cavity rods. If mortar fragments remain, they may lodge on wall ties lower down the wall or collect near the damp-proof course, where they may be retained by slight projections of the dpc into the wall cavity.

A reasonably well-ventilated cavity will dry any bridged mortar ties more quickly than one that is sealed. Weephole outlets should be provided to discharge any water that does collect in the wall cavity, particularly near damp-proof courses or cavity trays. Thus weepholes should be below dpc's and above cavity trays. Cavity trays are necessary to protect the heads of any openings associated with windows or doors, and care should be taken to ensure that any horizontal projections into the cavity do not provide a route for dampness (see Chapter 8).

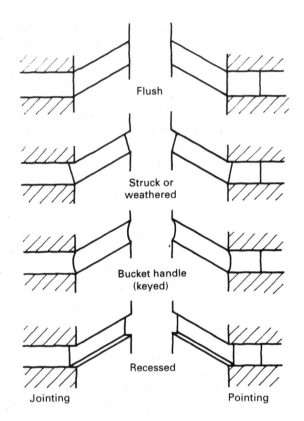

Fig. 5.7 Joint and mortar pointing profiles: bucket handled and struck (weathered) are best for resisting rainwater penetration (BS 5628 Part 3).

Another problem with rainwater penetration in cavity walls is its effects on wall ties. Metal wall ties, which are by far the most common type, should be protected against corrosion. They should never slope inwards, as this will enable water from the damp outer skin of brickwork to cross to the inner leaf. They should have a drip on the centre of the tie to prevent this water movement.

Research at the BRE suggests that corrosion of galvanised mild steel wall ties is probable in all houses with cavity walls built prior to 1981 (BRE Digest 329). The rate of loss of galvanising on metal wall ties has been estimated at $15\,\text{g/m}^2$ per annum. Butterfly ties had galvanising up to $260\,\text{g/m}^2$ and $460\,\text{g/m}^2$ on metal twist ties, but many ties of 'commercial' quality with as little as $50\,\text{g/m}^2$ were used up to 1981, when a new amendment to BS 1243: 1978 requiring $940\,\text{g/m}^2$ galvanising was published. It is predicted that failure of wall ties may become increasingly frequent in houses built with cavity walls during this post-war period. In addition, cavity wall dwellings within a few kilometres of the sea, or built using black ash mortar, will be susceptible to early failure because of the accelerated corrosion induced by these influences.

Wall tie corrosion is most likely in the external leaf, and the consequent expansion may cause horizontal cracking along brickwork courses at tie positions (i.e. every four courses). In severe cases this expansion will cause the outer leaf to bow noticeably. Such cracking will, of course, aggravate any further rainwater penetration. If such corrosion occurs at the inner leaf, floor and roof lifting, shear cracking on bonded walls and wall disruption may occur. At the worst, corrosion will occur on the embedded sections, and thus borescope inspection may not reveal the full extent of the deterioration. The only sure way of confirming the corrosion may be to remove some bricks to inspect the ties and the thickness of galvanising. As butterfly ties have less bulk, any effects of corrosion may be even less obvious without opening up.

Remedial measures in the event of wall tie failure include the drastic expedient of demolishing the outside leaf, which is usually non-loadbearing, and rebuilding it after removing all the original ties (BRE Digest 329). Alternatively, the ties could be individually located using a metal detector and replaced by the removal of a few bricks at each tie position, but this would be a slow and laborious task, which would be very difficult to achieve neatly.

Special stainless-steel wall panel restraint ties have been developed by Hilti (BBA certificate 89/2713) and Redhead, to name but two manufacturers, which may be used where the outer leaf can be saved. These can be installed by drilling holes into the wall in which the ties are inserted, near but not at the same position as the existing and at the same distribution. After insertion, either mechanical expansion bolts or bolts set in resin provide a firm joint in the masonry. The bolt holes are filled with grout afterwards. The installation of structural polyurethane foam, which would also improve the thermal insulation, is another alternative remedial option, but its long-term performance is still uncertain.

Claddings

The most common modern form of walling for a building envelope is cladding. It is usually a non-load-bearing element designed only to sustain its own dead weight and resist wind forces. With the exception of some reinforced concrete panels, such claddings are not designed to take structural loads. The outer leaf of cavity walling is of course a form of cladding. The other main forms of modern cladding are listed below, all of which with the exception of some precast systems can be classed as thin claddings:

❑ curtain walling;
❑ stone facings;
❑ profiled metal/asbestos cement/plastic sheeting;
❑ GRP/GRC panels;
❑ stone facings;
❑ precast concrete panels;
❑ rainscreen/raincoat claddings (see Chapter 8).

The main problems associated with most types of cladding are:

❑ air/moisture ingress/leakage;
❑ interstitial condensation;
❑ corrosion of fixings;
❑ delamination or deterioration of the face or coating;
❑ fire spread behind the facing.

The primary water ingress problems with claddings are dealt with in more detail in Chapter 8.

Parapet walls

A parapet wall, like a chimney, is exposed to a greater degree of weather exposure (Fig. 1.1) than almost any other part of a building. Thus as a projection above the roof it is a source of potential weakness so far as the exclusion of rainwater is concerned. Many parapet walls were designed and built well before the introduction of building regulations, and some of the details now considered essential for the exclusion of water were not adopted.

The top of the parapet should be finished with a coping (BS 5624, 1983) rather than with bricks on edge, which are very vulnerable to frost attack and leakage at the exposed joints. In any event, the coping should be of an impervious stone or precast concrete, under which must be a dpc to prevent water penetration at the joints. The recommended 40 mm minimum projection of the coping on either side should be designed with an adequate throating to 'throw clear' any rainwater at this position from the supporting masonry. In higher parapets, a second dpc will be necessary above the roof line to prevent any dampness that has penetrated through the outside wall from decaying adjacent roof timbers or penetrating to spoil internal decorations. This lower damp-proof course should be continuous with the valley gutter behind the parapet wall (Fig. 5.8).

In modern buildings, parapet walls are usually of cavity wall construction. They should have twice as many vertical expansion joints as

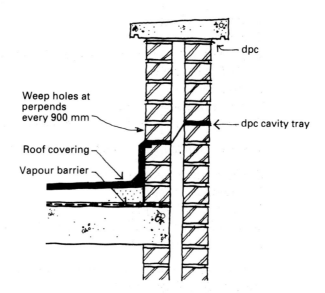

dpc

Weep holes at
perpends
every 900 mm

dpc cavity tray

Roof covering

Vapour barrier

Fig. 5.8 Parapet wall construction. A 40 mm overhang with a throating is recommended, together with a dpc below the coping stone. In higher parapets, a second dpc will be necessary above the roof line, which is continuous with the waterproofing layer of a valley gutter, or roof line (BS 5628 Part 3).

the main walling below them. In older buildings, solid parapet walls are to be expected. Penetrating dampness through the external facing of such parapet walls can be hazardous to any adjacent roof, framing, wallplate, bonding or joinery timbers.

Walls can be protected by projections such as eaves, cornices or string courses, but these should be constructed from impervious materials, or covered with impervious coverings such as lead (very appropriate for older buildings) or asphalt, polyester felt or strip metal flashings (Fig. 5.9). These may have an effect on water flow down the front of the building, and if they are incorrectly designed or develop faults, they may have an adverse influence on the water flow, and this may lead to moisture penetration.

External joinery

External joinery has traditionally been constructed of timber. Today aluminium is still being used, and uPVC (unplasticised polyvinyl chloride) plastic, with its promise of durability and low maintenance, is an attractive alternative. It remains to be determined whether this promise is realised over the life expectancy of the material. Indications

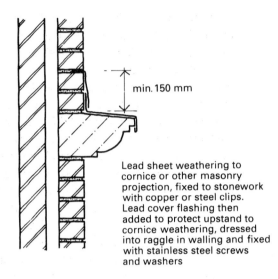

min. 150 mm

Lead sheet weathering to
cornice or other masonry
projection, fixed to stonework
with copper or steel clips.
Lead cover flashing then
added to protect upstand to
cornice weathering, dressed
into raggle in walling and fixed
with stainless steel screws
and washers

Fig. 5.9 Flashings for projections.

are that plastic external building products have a minimum service life of
at least 20 years. Plastics, however, are susceptible to problems such as
embrittlement and bleaching of colour as a result of excessive ultraviolet
light exposure. Anti-yellowing ingredients in the plastic have generally
minimised this problem. Neither uPVC nor aluminium windows, however, are as repairable as their timber counterparts.

External joinery commonly suffers from problems associated with
lack of durability and poor maintenance. The greatest concentration of
these is found on post-war buildings constructed prior to the 1970s, when
preservative-treated timber was introduced (BRE Digest 321).

Up to early Victorian times, most of the external joinery used on
quality buildings that survive today would have been made from oak
from which the perishable sapwood would have been discarded. The
durability of oak heartwood is well known, and it would perform very
well on external weather exposure. When softwoods were substituted,
Baltic redwood from Scandinavia and Russia was the chosen species.
The heartwood of such timber had a reasonable degree of durability.
However, the sapwood proportions in those days would have been small
(Fig. 4.1), and would not have featured in a large proportion of the
joinery timber used. Gradually, as virgin forests were felled, younger
second-grade trees would have been used, with relatively more sapwood
and greater susceptibility to decay.

In the Victorian period, most joinery would be handmade, and joints
would have been well primed with a lead-based priming paint. Priming

would have been applied to end-grain areas, to the backs of components, and to the surfaces of glazed rebates as well. Modern mass production does not always allow for end-sealing and back-priming. Furthermore, in recent years lead-based paints have been replaced on environmental grounds by non-toxic alternatives. The avoidance of lead paints also removed a degree of fungicidal protection from the film, allowing staining moulds to become established more readily (see Chapter 3).

At the same time, solvent-based systems have been replaced with quicker-drying and cheaper emulsion products from several joinery manufacturers. This substitution by the more porous emulsion paints would allow moisture pick-up and loss by the timber to a greater extent than was required with the more traditional oleoresinous paints. Thus the absence or low standard of priming would allow timber moisture content to increase rapidly during wet weather. The joints would open up, especially on the bottom rails, as the result of differential moisture movement of the timber, and water penetration would occur. Furthermore, the site practice of cutting off window 'horns' to facilitate building into the surrounding brickwork would expose the unprotected end-grain of the timber. Much joinery is also exposed to inclement weather conditions on the building site during construction work, leading to moisture content increases. High moisture contents will mean a greater likelihood of wet rot decay in such timber.

The absence of fungicidal protection of paint has permitted the development of various mould fungi on exposed wood surfaces. These non-rotting fungi can also grow through the paint film, and they can develop in the underlying timber, especially if it is unprotected by paint and at a higher moisture content. Although they do not actually decay the wood, they live on the nutrients deposited in some of the wood cells. Their main influence, however, is to rupture the paint film by forming pustules of spore just under its surface. These 'punctures' allow water penetration to occur. This is so-called 'blue stain in service' (see Chapter 4).

The more widespread introduction of central heating in buildings in the 1960s has resulted in greater condensation on external single glazing. Such condensed water would naturally fall to the bottom of the glass and infiltrate the bottom of the sill or sash component, leading to higher moisture content, especially if priming is of a poor standard or with a porous paint. Even in double-glazed components, condensation on the glass is still possible, as it will usually represent the coldest surfaces close to the high internal humidity conditions.

As a result of the NHBC initiative in introducing a treatment specification, since the 1970s there has been an almost universal adoption of preservation treatment of external joinery. Normally organic solvent-type preservatives are used, applied by the immersion or double vacuum methods. This seems to have controlled decay in such treated external joinery.

More adequate and correct application of good priming paint materials on the joinery and better site protection of components are regarded as essential prerequisites to better paint and coating performance on external joinery. In the long run, regular repainting (say, every four or five years) is required for external joinery to ensure that it is adequately maintained. Other treatments, such as the incorporation of water-repellent wood preservatives and the use of external stains and microporous paints, have been discussed in Chapter 3.

Metal, aluminium and replacement window joinery

In the immediate pre- and post-war periods, hot-dip-galvanised metal windows were very popular. In practice, these windows suffered from many problems. The galvanised coating did not prevent rusting, and poor insulation properties attracted condensation. They were frequently draughty, owing to the difficulty of achieving a good fit with the opening lights. Generally, timber windows were cheaper and more appealing, and consequently the use of galvanised metal frames declined.

Aluminium windows became popular in the 1970s, especially as replacements for older wooden frames that had either decayed or needed too-frequent repainting maintenance, or just as part of the general improvement of the property. Improved function – double glazing, draughtproofing, better durability and low maintenance – is perhaps more important in replacement than in some new construction, where initial cost is still an overriding issue.

Aluminium frames, however, do suffer from being poor insulators. Condensation on the frames of such double-glazed windows can only be reduced, not always eliminated, by designing a 'thermal break' in the frames to prevent a 'cold bridging' effect. This can be done by incorporating a plastic foam within the voids of the frame, and using hardwood subframes. Double-glazed polyester-coated aluminium or steel frames may perform in a similar way. Aluminium frames are also susceptible to alkaline corrosion if placed next to concrete or other cement-based materials.

Window frames of uPVC have become widely used, not only for replacement purposes, but also in new build. Even with their higher initial costs, uPVC frames may be acceptable because of their superior insulating properties and apparent lower maintenance costs. Indeed, one study indicated that in total window cost terms uPVC was best, followed by softwood, then aluminium, with hardwood last (Essex, 1993). However, their lack of strength, the tendency of bleaching or yellowing of the plastic, the difficulty of repair and lower flexibility in coping with the wide temperature ranges in the UK may affect their long-term success.

Hardwood window frames are also used for replacement windows. Hardwoods are selected because of their expected durability and decay resistance, and because of their attractive appearance when used with a clear or translucent finish. It is important to ensure that such hardwood, if used untreated, does not contain any sapwood and is of a sufficiently durable species (e.g. mahogany rather than obeche). All wood frames do have the great advantage of excellent insulation properties, which is so useful in periods of condensation, and they are relatively easy to repair.

Any window replacement scheme should comply with the requirements of BS 8213: 1990 as regards cleaning and safety. On no account should dead/fixed-pane windows with top hopper lights without sashes be used on any storey above ground-floor level. Trickle vents should be incorporated in the head of each window frame (see Chapter 6).

Doors

Generally, the performance of external doors is much the same as that of windows. Timber doors and frames require priming protection on all surfaces, and this includes the underside of the doors and the ends of the framing members, which may be cross-cut and fitted on site, leaving unprotected end-grain exposed. Many doors are manufactured using untreated perishable timbers as mouldings and dowels. Ramin and obeche are common choices, and these may act as a focus for decay if and when moisture penetration occurs. Decay can also develop if non-durable plywood is used on external doors.

As with all external timber joinery, joints and end-grain areas can be vulnerable to water penetration. It is therefore the design of such components, the selection of appropriate timber species, and suitable preservative and protection treatments that must be carefully considered to achieve satisfactory performance.

See Chapter 8 for further consideration of the avoidance of water penetration problems at doorways.

Timber frame systems

Traditional timber frame

Medieval timber frame construction was based on large-sized timber framework, usually of oak, and walls were 'infilled' with 'wattle and daub' (RICS Building Conservation Group, 1993). Wattle could be typically of split oak and chestnut staves and hazel twigs intertwined, covered with clay, lime plastered and given a period decorative coating of 'distemper' or whitewash (i.e. limewash). The external surface of the timber framework would be exposed to the weather, and over the centuries often deteriorated severely.

In more recent times, the wattle and daub infill was often replaced with a single skin of brickwork. This is very undesirable for several reasons. Water penetration could occur through the single skin of brickwork and could cause decay in the adjoining timber. The wall would have a greatly inferior insulation level compared with wattle and daub, and the building would become cold and more expensive to heat. Finally, the mass of the brickwork would be much greater than that of the original infill, and this weight could be excessive, especially in overhanging jetties.

If the timber framework becomes persistently damp, as frequently happens near ground level in such buildings, the timber gradually decays. The development and extent of the decay is easily detected, and local treatment and repair using special systems such as epoxy resin cement filler products and pressure injection of siliconate preservatives can often be adopted. Such localised repairs can be much cheaper and easier to do than total renewal of the defective timbers, provided the decay is not extensive.

Non-traditional timber frame

Some timber frame buildings, which are neither the 'period' nor 'modern' type, were built as part of the non-traditional housing phases from the early 1900s to the late 1950s (Ridley, 1988). These were mainly low-rise buildings with slated or tiled pitched roofs, and were often built using the balloon frame method of construction. They can sometimes be easily identified by the distinct timber weatherboarding exterior finish. Defects such as wet rot to the claddings and sole plates account for the more serious moisture-related problems in such buildings.

Modern timber frame

The demise of discredited precast concrete system construction for housing by the late 1960s encouraged a shift towards timber frame. Since the early 1970s there has been a considerable increase in the number of houses built to the modern timber frame system. The platform frame system is the most common method of this type. In the mid-1980s, following unfavourable comment in the TV and press, there was a significant reduction in timber frame construction output. By the early 1990s, however, the timber frame industry made a recovery, by tightening up its on-site procedures, and regained a large part of the housing construction market.

Most of the adverse publicity from the media has been directed towards inadequate site supervision leading to incorrect construction. One of the main results of such poor construction was that incorrectly built cavities and inadequate or improper vapour checks led to high moisture levels in the timber frame. These deficiencies not only reduced the effectiveness of the insulation, but also allowed rot to develop.

The modern timber frame system mainly adopted in the UK has an internal framework of prefabricated wall panels consisting of timber stud and plywood sheathing, which supports the roof and finishes. Internal stud partitions can be of similar construction, and frequently are load-bearing as well. To improve insulation, the 100 mm wide space between the timber studs in the inside frame panel is filled with mineral wool insulation, and to prevent interstitial condensation in such a structure a vapour check is placed behind the internal plasterboard lining. Every attempt is made in the design and construction to ensure that this vapour check is as continuous as possible, and novel techniques to protect the cavities behind pipes, electrical sockets and switches have been developed.

As in other forms of cavity construction, the cavity is designed to protect the inner leaf, in this case the timber framework, from moisture penetration from the outer brickwork skin or around window and door openings, for example. The use of a damp-proof course and a waterproof but vapour permeable 'breather' (or moisture) barrier (sometimes referred to as 'building paper') protects the frame from penetrating dampness in the cavity (BS 4016: 1972). At the same time it prevents any vapour that has penetrated the inner vapour check behind the plasterboard from being trapped and condensed in the insulation space between the timber studs (Fig. 5.10).

Provision has to be made for greater moisture shrinkage of the inner timber frame as compared with the brick 'outer veneer'. Care must be

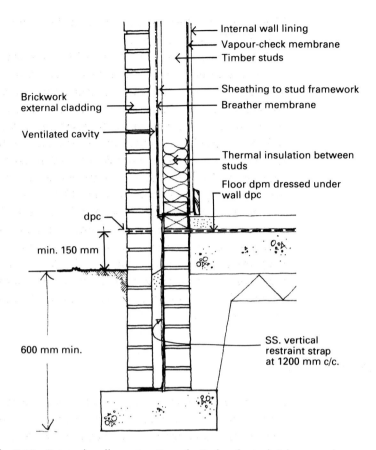

Fig. 5.10   External wall construction of a timber frame brick veneer house.

taken to ensure that, as with other cavity construction, the main cavity is not blocked with mortar droppings, and that the wall ties are correctly installed and do not act as channels for penetrating damp. It is also important to ensure that the main cavity spaces are adequately ventilated, because they may be blocked at regular intervals (e.g. 5 m horizontally and every storey vertically) by cavity fire stops.

It is generally accepted practice that both the sole plates and the timber framework are preservative treated. In addition, the plywood sheathing should also contain preservative and fire-retardant treatment using borates.

Several surveys have been conducted by the BRE to try to establish the likelihood of decay developing in timber frame houses (Covington *et al.*, 1992). The general indications from such investigations suggest that timber moisture contents in the frame are not excessive or dangerously high even soon after construction, and that there is a tendency for the timber to dry out with time.

Data from the NHBC in the late 1980s showed that 150 000 timber frame houses had been completed and are covered by their ten year warranty. The rate of claim on such timber frame houses is 1 : 3000 compared with traditionally built (i.e. masonry) houses with a claims record of 1 : 79. The majority of claims relate to incorrect installation of wall ties or insufficient provision of movement of the timber frame on initial drying. It is this favourable record that has contributed to the re-emergence of timber frame construction as a popular house-building method after the bad publicity of the 1980s.

From the theoretical standpoint, it is probable that the temperature fluctuations that will occur in the lightweight timber wall frame would be over a range that, even if moisture conditions favoured dry rot, the temperature would not. If wet rot develops as a result of a leak in plumbing or central heating, for example, the renewal of a small area of decayed timber is much easier and quicker than many other building repair operations.

Damp-proof courses

Most building materials are porous to some extent, and damp-proof courses are therefore essential to prevent moisture from penetrating and reaching the habitable parts of a building, or affecting building materials that would be seriously damaged by water.

A damp-proof course is intended to provide a barrier to the passage of water from the exterior of the building to the interior (BS 8215: 1991). There are a range of materials used, and their properties are summarised in Table 5.3. A damp-proof course should be constructed from a material that has a life expectancy equal to that of the building, and it should resist compression and sliding. Dpc's should extend to the full thickness of the wall or leaf, and should project beyond the external face. They should be laid on a fresh bed of mortar, which does not contain any coarse aggregate. Dpc's are usually polythene based (BS 6515: 1984) or bitumen based (BS 6398: 1983).

A damp-proof course should be at least 150 mm above outside ground level in external walls. This distance is generally considered to be high enough to prevent water splash-up affecting the wall above the dpc. The damp-proof course in the wall must be continuous with any damp-proof membrane (dpm) in the floor (Fig. 5.11).

Where a damp-proof course is inserted in an external cavity wall, the

Table 5.3 Properties of materials for damp-proof courses (BS 5628 Part 3: 1985).

Type	Minimum thickness (mm)	Joints	Risk of extrusion	Durability	Remarks
Flexible					
Lead to BS 1178	1.8	Lapped and/or welted	Unlikely	Corrodes on contact with lime and cement	Protect with bituminous paint
Copper to BS 2870	0.25	Lapped and/or welted or welded	Unlikely	Can corrode on contact with salts	Protect with bituminous paint
Black polythene 0.915–0.925 g/l to BS 6515[a]	0.46	Lapped and/or welted	Unlikely	Good	Accommodates lateral movement
Bitumen felt BS 6398 Hessian, fibre or asbestos base		Lapped and/or sealed	Can extrude under heavy pressure or heat	Good	Warm before laying in cold weather
Bitumen and pitch polymer	1.10	Lapped and/or sealed	Unlikely	Good	Accommodates lateral movement
Semi-rigid					
Mastic asphalt 12 mm to BS 1097 and 6577			Can extrude under heavy pressure	Good	Score or grit for good key
Rigid					
15% epoxy resin/sand	6.0		Not extruded	Good	
Dpc bricks BS 3291	Two courses; use 1 : 3 cement sand		Not extruded	Good	For rising damp only
Slate to BS 743	Two courses; use 1 : 3 cement sand		Not extruded	Good	

[a] Polythene does not form a good bond with mortar and so is not suitable for use where compressive strength is low, i.e. under copings.

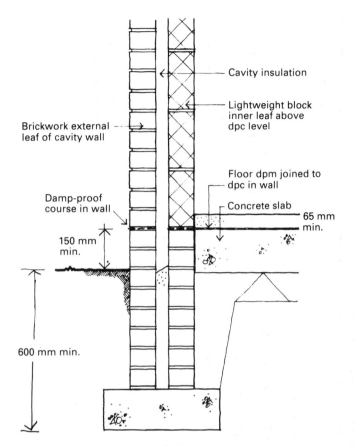

Cavity insulation

Lightweight block
inner leaf above
dpc level

Brickwork external
leaf of cavity wall

Floor dpm joined to
dpc in wall

Concrete slab

Damp-proof
course in wall

65 mm
min.

150 mm
min.

600 mm min.

Fig. 5.11   Damp-proof courses in solid floors.

cavity should extend at least 150 mm below the lowest level of the dpc. Where a cavity wall is built off a raft foundation or similar supporting structure, the supporting structure should be regarded as bridging the cavity, and should be protected by a cavity tray dpc.

Solid ground floors

In medieval buildings, all solid ground floors consisted of packed earth, on which were laid natural stone flags, bricks or tiles as a floor finish. No dpm was provided, as these floor finishes were probably selected in an attempt to provide a barrier to ground dampness. Today, in-situ concrete is the universal choice for solid ground-supported floors, with various materials used as the floor finish.

Building regulations require that a ground floor is constructed so that moisture is prevented from passing to the upper surface of the floor. These requirements are usually met by laying dense concrete incorporating a dpm on blinding on a hardcore bed. The hardcore consists of large and coarser particles, which should reduce the level and extent of capillary rise of ground water. The material selected for hardcore (preferably Type 1 grade crushed brick or whinstone) should be hard, durable and chemically inert, and free from water-soluble sulphates or other deleterious matter, which might cause damage to the floor. The function of the hardcore is to level the site, to raise levels in respect to the surrounding ground levels, and to provide a firm, earth-free base for the concrete slab.

Polyethylene (Polythene) membranes may be laid above or below the concrete floor slab, with their joints sealed (BBA 94/3068). The Building Regulations state that a minimum 1000 gauge (i.e. 0.25 mm thick) polyethylene should be used. Nowadays 1200 gauge (i.e. 0.3 mm thick) Visqueen or other similar polythene sheeting should be used, as this thickness is better at resisting damage from puncturing. In more severe cases 2000 gauge (i.e. 0.5 mm thick) is recommended (Table 5.4).

Table 5.4   Performance of waterproof flooring materials.

Type of finish or underlay	Performance
Mastic asphalt to CP 204	Impervious to moisture in the liquid or vapour form
Pitchmastic to CP 204	
'Sandwich materials'	
Mastic asphalt to BS 6577	Impervious
Bitumen sheet dpm to BS 743	If the joints are sealed, impervious
Hot pitch or bitumen	Impervious at 3 mm thickness
Cold-applied bitumen solution and coal tar pitch/rubber or bitumen rubber emulsion	Impervious if applied by repeated brush application to achieve adequate thickness
Polyethylene	If laid with 150 mm laps and tape-sealed joints with no punctures, is satisfactory under some floor finishes

If the dpm is laid below the slab, it should be supported on a blinding layer of whin dust on a 150 mm thick bed of hardcore. This is to be preferred to the other method, in which the dpm is placed above the slab. Although this latter method obviates the need for a layer of blinding, it inevitably leaves the concrete slab open to penetration by ground moisture, which might contain deleterious substances such as sulphates and acids. In such cases sulphate-resisting cement should be used in the concrete mix.

In cases where the dpm is laid above the slab, a number of materials may be used other than polythene sheeting: mastic asphalt, bitumen sheet, hot-applied pitch or bitumen, cold-applied bitumen (e.g. Aquatex or Synthapruf), coal-tar pitch or bitumen rubber emulsions. The floor screed and finish are then laid. A floor screed laid directly onto a dpm will in effect be a separate layer, and thus should be at least 65 mm thick to avoid curling and other problems with the screed.

Most flooring materials are affected by dampness, although a few will transmit moisture without failure. Materials such as magnesium oxy-chloride (otherwise known as 'magnesite' or 'composition' flooring) are particularly susceptible to moisture. Disintegration starting on the surface is usually due to inefficient maintenance of the floor finish so that spilt fluids or cleaning water are not cleaned up properly (PSA, 1989). Also, a defective dpm will cause similar problems in such flooring.

The performance of membranes and floor finishes is summarised in Table 5.5.

Table 5.5   Floor finishes and ground water penetration (BS CP 102: 1973).

Type of finish	Performance
Pitchmastic and mastic asphalt flooring	Resists rising damp and is unaffected by it
Concrete Terrazzo Concrete or ceramic tiles	Transmit dampness without material failure, unless exposed to sub-zero conditions, which may result in frost damage
Cement/latex Cement/bitumen	Partial transmission of dampness without material failure and generally without adhesion failure
Thermoplastic tiles (BS 2592) Vinyl asbestos tiles (BS 3260) Magnesite (BS 776)	Partial transmission of dampness, especially at joints
Flexible PVC as a sheet or tile (BS 3261) PVA emulsion cement Rubber	These finishes all need reliable protection against dampness either because they are directly affected by it or because the adhesion is sensitive to moisture
Linoleum Cork/wood/chipboard	Polythene membranes may not be satisfactory to protect these last two groups

CP 102: 1973 and the Building Regulations state that the slab should be at least 100 mm thick. However, in our view, 150 mm is a safer and more realistic minimum thickness for solid ground-supported floor slabs. Quality control on construction sites can be very difficult to achieve consistently. This increased thickness is an example of the 'creative pessimism' propounded by Addleson & Rice (1991). It is not to say that a slab 100 mm thick will be inadequate. Rather, the increased thickness allows for possible discrepancies in the depth of the concrete, which may occur at the time of laying. In some situations there may be a tendency by site operatives to undersize rather than oversize the floor slab because of shortages of available concrete, and other pressures. Thus a 100 mm specified floor slab may actually be only 95 mm thick once installed. Such deficiencies will inevitably reduce the moisture resistance and load-bearing capacity of the floor slab.

If it is proposed to lay a timber floor finish such as parquet or other wood blocks directly onto concrete, the timber can be bedded in a material such as bitumen, which may also act as a crude moisture barrier. However, any such adhesive should not be relied upon to act as a proper dpm. If a timber floor finish such as tongued and grooved boarding is being used, it can be fixed either to preservative-treated wooden battens laid on concrete or to special stainless steel clips. Neoprene foam strips can be placed under the battens to enhance the floor's acoustic performance. This is to be preferred to fixing the boarding to battens embedded in the concrete.

In older buildings, particularly in basement areas, timber floor finishes laid onto a solid concrete or stone floor slab, perhaps on a bed of cinders or fixed onto battens laid on the substrate, are often encountered. Such floor constructions are highly vulnerable to either penetrating damp or interstitial condensation, and as such would not be approved under current regulations. It is impossible to bring these floors up to modern standards without relaying. Surveyors should therefore be wary when this form of construction is encountered in a basement situation.

Suspended ground floors

General

Suspended ground floors are normally constructed either from timber, which is the traditional method, or from precast concrete, which is the more modern system.

Whatever system is used, however, subfloor ventilation is critical in all suspended ground floors, for the following reasons:

❏ to allow any evaporating moisture to escape to the outside air;
❏ to prevent the build-up of stagnant air;
❏ to avoid interstitial condensation;
❏ to prevent the build-up of hazardous subsoil gases such as radon and methane;
❏ to allow any leaked gases from pipework or appliances to escape.

Suspended ground floors have several advantages. They can usually be laid much quicker than solid floor constructions. The use of land for building which in the past might have been considered unsuitable or uneconomic for building, such as sloping sites, landfill areas or other 'marginal' land, also favours suspended floors. Suspended floors are claimed to be less liable to being affected by ground heave or clay shrinkage. Easier laying and connection of services and 'dry' construction are other advantages. In any event, a hatch or trap door 600 × 600 mm should be provided in all suspended floors for access to the subfloor void for maintenance and services installation work.

Suspended timber floors

These usually consist of timber tongued and grooved floor boards nailed to floor joists, which are supported on timber wallplates, which in turn are supported on either a scarcement, or on brick sleeper walls built off the oversite concrete, or on joist hangers (Fig. 5.12).

Where sleeper walls are used, they should be constructed on 150 mm rather than 100 mm thick oversite concrete, again to allow for any possible deficiencies on site. Although no actual dpm was mandatory under the regulations for suspended floors, in Scotland for example the oversite concrete must be finished with a two-coat application of hot-applied bitumen. This 'solum treatment' helps to reduce moisture evaporation from the slab and prevents the growth of vegetable matter in the subfloor void. Nowadays it is advisable to include a 1000 gauge polythene membrane below the oversite concrete instead. Needless to say, the solum area should be cleared of rubbish, wood shavings and other builder's waste. Such debris can act as a ready reservoir for moisture. Unfortunately, however, subfloor areas are often left uncleared. This should be noted by the surveyor during a subfloor inspection.

In many buildings constructed prior to the 1960s the concrete and

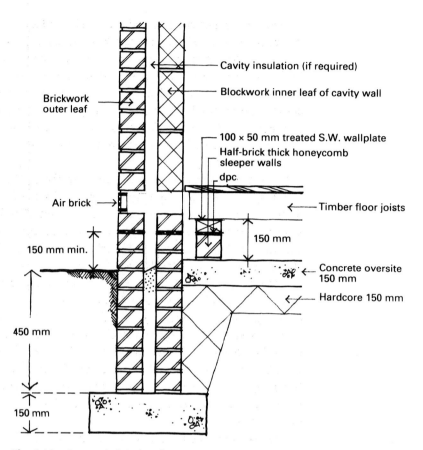

Cavity insulation (if required)

Blockwork inner leaf of cavity wall

Brickwork
outer leaf

100 × 50 mm treated S.W. wallplate

Half-brick thick honeycomb
sleeper walls

dpc

Air brick →

Timber floor joists

150 mm

150 mm min.

Concrete oversite
150 mm

Hardcore 150 mm

450 mm

150 mm

Fig. 5.12 Suspended timber floor construction.

hardcore oversite is uncommon, and in many older buildings the solum finish is earth. Bare earth oversite finishes allow subsoil moisture to evaporate into the subfloor void, which in winter periods could increase the moisture content of joists and other timbers above the 18% safe threshold.

It is very desirable therefore to incorporate a layer of polythene over such bare earth oversite as part of building rehabilitation work, to prevent such evaporation and to assist in the drying-out of the subfloor space. The height of the subfloor space should extend to at least 125 mm below any suspended timber and a minimum of 75 mm below any wallplate.

Timber must be protected from ground dampness by the use of a physical damp-proof course, and the subfloor space must be adequately ventilated by the provision of satisfactory through-ventilation (BRE Digest 364). The sleeper walls (also known as 'dwarf walls') must be of honeycomb construction, allowing a free flow of air under the floor from the fresh air inlets (FAIs) or airbricks in the external walls. FAIs should provide 3000 mm^2 open area per metre run of external wall (Part C, 1991 Building Regulations). Single brick-size FAIs are usually at 1.2 m centres. As has been pointed out by Harris (1995), however, subfloor or crawl space ventilation requirements appear to have developed in an ad hoc manner with no technical basis. More research is clearly needed in this area, particularly with regard to long subfloor spaces with and without recesses.

Unventilated pockets or 'dead zones' in the subfloor space must be avoided, but these are often encountered in floors of old buildings with 'L-shaped' rooms or rooms with large recesses. Another problem can arise where a suspended floor joins a solid floor, as may happen when an extension is built. In such cases, special ventilating ducts of at least 100 mm in diameter under the solid floor area will usually be necessary to link the blocked-off subfloor void with the external wall.

Ideally, the solum level should be the same as or higher than the outside ground level. In many old buildings, however, this is not the case. The problem with a low solum level is that it can be difficult to ventilate properly. This is usually achieved by forming a small well in front of the FAI, but these are all too often filled in with earth. Also, in the event of flooding, the solum can act as a reservoir for water. This water will provide a ready source of evaporated or liquid moisture for fungal attack to the exposed timbers.

The use of cranked and ducted air vents (Fig. 5.13) to ventilate subfloor spaces where a suspended floor is close to external ground level is a recent development. These ducted vents offer a possible solution to a frequently encountered problem that may otherwise necessitate lowering the outside ground levels, or replacement by solid floor construction.

Frequently, suspended timber ground floors in older buildings are not well ventilated. Cast-iron vent grilles were the most common type of vent cover in such properties. Either insufficient FAIs were installed during the construction, or vents became blocked by raising outside ground levels or through later building repair and alteration work. In such situations, polypropylene fresh air inlets are useful in that they allow greater air flow than is possible with the same size terracotta air brick.

Joists should always be kept clear of external walls, but in older constructions they may be supported by a scarcement or in the inner layer of such walls, or even built into solid 225 mm thick construction. The ends of built-in timbers are always at risk from decay. It is also quite usual to find no damp-proof course below wall plates and the floor plate

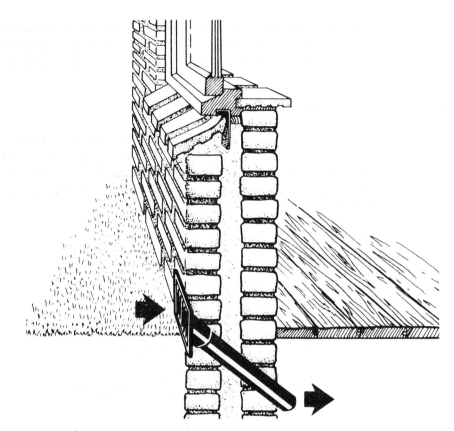

Fig. 5.13 Ducted air vent to improve subfloor ventilation where there are high external ground levels (UBBINK, UK Ltd).

timbers, and it is quite surprising how such floors survive for so long. The isolation of timber from damp masonry is always a desirable aim. One way of doing this when rebuilding floors where decay has occurred, for example, is to support the joists on galvanised or stainless steel hangers rather than reinsert them into the masonry recesses. If this is unavoidable, the joists should be inserted into an impervious glove or 'shoe box' before slotting them into the pocket formed in the masonry. It is also important to isolate timber floors, where possible, from solid floors, which in older construction were often laid without any dpm.

Shrinkage of square-edged or butt-jointed floorboards (which are sometimes found in very old properties) can allow a draught to develop from the subfloor ventilation. This is less of a problem when tongued and grooved boards are used, and is virtually eliminated when chipboard sheet floors are laid. Chipboard can perform well if supported properly provided it remains dry, but the normal grades are very susceptible to dampness, and rapidly lose strength when they become wet. A moisture-resistant grade (type 3 – BS 5669) is available, and this should always be used if there is a risk of dampness, in bathrooms or kitchens for example.

Suspended precast concrete floors

In many residential building work, the use of suspended concrete floors is now very popular. Such floors are potentially more durable and have better load-bearing properties than timber floors. Suspended precast floors consist of either prestressed concrete planks or prestressed concrete beams with infill blocks (Fig. 5.14).

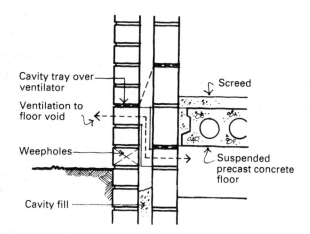

Fig. 5.14 Suspended precast concrete floor construction.

In some previous installations the provision of subfloor ventilation or solum treatment was never a requirement. For the reasons stated earlier, however, it is strongly advisable to ventilate such large void areas and apply oversite concrete to the bare solum. In addition, it is now considered good practice to provide some form of ground cover in the solum areas of all suspended ground-floor constructions with a vapour-resistant material such as polythene sheeting (Rose, 1994). Ventilation can be achieved by using stepped ventilators with a cavity tray above each one (Tovey & Roberts, 1990).

Chapter 6
Condensation

Background

Condensation can reasonably be described as a modern disease of buildings. It is probably fair to say that it is the most significant form of dampness in the UK. The increase in the incidence of condensation in British buildings is due to many changes in relation to both construction standards and domestic living habits in the second half of the twentieth century (IMBM, 1986).

The various national house condition surveys carried out in the UK since the 1960s have clearly indicated that millions of households are affected to a greater or lesser extent by condensation. In some instances these problems have been so severe and have so defied rectification that demolition has been the ultimate remedy.

What is condensation?

Condensation is the formation of extra water vapour in the form of a liquid when saturation of the air occurs (BRE Digest 110). Moisture in the form of water vapour is present in our external environment at all times. In the UK the levels of atmospheric water vapour are high compared with many other parts of the world. This is because of its northern latitude and the fact that the prevailing westerly air flow picks up water vapour during its long passage over the Atlantic Ocean.

The amount of water that can be present in the air is related to the ambient temperature: warmer air can contain more water vapour than cooler air. When the upper limit of water vapour is reached, the air is said to be saturated. The amount of water vapour that can be contained by saturated air is dependent on the temperature of that air (Table 6.1).

The actual amount of water vapour in the air is expressed as humidity relative to the maximum saturation level. This is known as relative humidity, or RH. If the temperature falls in an already saturated air

Table 6.1   Water vapour content of saturated air (100% relative humidity).

Air temperature (°C)	Water vapour content (g/kg air)	Vapour pressure (mbar)
5	5.3	8.7
10	7.5	12
14	10	16
18	13	21
24	18.75	30.4

(100% RH) where the air is said to be at its dew-point, the fall in temperature results in some of the water vapour being condensed. The amount of condensation water produced will depend on the amount of vapour in excess of 100% RH. Thus by reference to Table 6.1, a reduction of temperature from 18 to 14 °C would condense 3 g of water vapour per kg of air. The term 'dew-point' describes the temperature at which condensation will occur in air with a specific water vapour content. As can be seen in Table 6.1, air with a water vapour content of 18.75 g/kg will cause condensation at any air temperature below 24 °C. The interrelationship of temperature, relative humidity and water vapour content can be seen on a psychrometric chart (Fig. 6.1).

Under normal conditions, water vapour in the atmosphere is in the form of invisible particles, which are suspended in the air. It is a gas, and exerts a pressure, which can be measured in millibars (Table 6.1, Fig. 6.1), and which increases with increasing water vapour content and/or increasing temperature. In occupied buildings most water vapour is generated in bathrooms and kitchens, and the pressure of this vapour will cause it to spread to all accessible parts of the building. This may, as a result, cause condensation in colder areas and in less well heated rooms. This diffusion of the humid air will occur most rapidly through doorways and stairwells, but ducts for plumbing or electrical services will also spread vapour passage. Condensation will therefore occur when this warmer vapour-laden air spreads to colder surfaces, which cause cooling of the air, leading to condensation of the excess water vapour.

Forms of condensation

Surface condensation (sweating)
Condensation can occur on many materials, but is most obvious on hard impervious surfaces such as glass, glazed tiles and gloss-painted plaster.

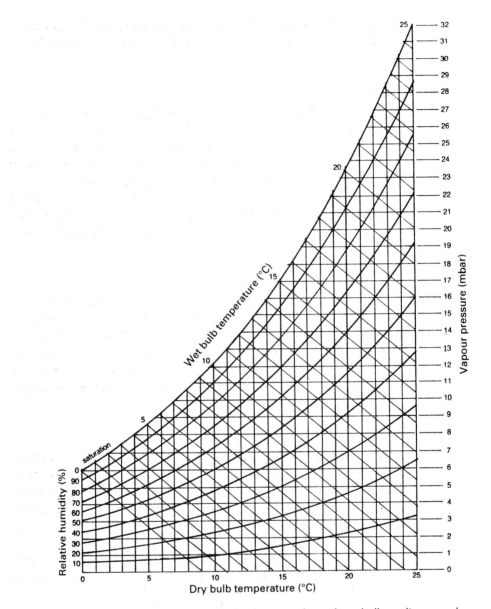

Fig. 6.1 Psychrometric chart. The relationship between dry and wet bulb readings on a hygrometer demonstrates relative humidity and vapour pressure. The chart can be used to calculate relative humidity (BRE Digest 110).

This is referred to as surface condensation (or sweating). In extreme cases it can easily be seen as beads of moisture on single glazing and other impervious surfaces. In many other instances, however, it is not readily visible, as porous surfaces such as wallpaper and fair-faced masonry will readily absorb the condensed moisture. Such forms of dampness may not be detected by visual or tactile senses (BRE Digest 297).

Interstitial condensation

Condensation that occurs within a structure is called interstitial condensation. It will thus not be immediately obvious where or when it occurs. It is only when its adverse effects occur, such as timber decay and deterioration of the insulation, that interstitial condensation will be detected (BRE Digest 369). There are, however, computer programs available that can be used to help predict the likelihood of interstitial condensation within building elements under standard steady state conditions (e.g. Owens Corning Ltd Architectural Calculation Suite, 1995).

Warm-front condensation

Condensation can occur sharply and unexpectedly on occasions if the external environmental conditions are right. If, after a cool spell, the weather suddenly warms up, dense elements such as floor slabs in a building may take several days longer to warm up to a temperature above dew-point than the internal air (Garratt & Nowak, 1991). This will cause the floor slab to sweat noticeably, and will soak any absorbent coverings such as carpets.

Reverse (or summer) condensation

This phenomenon can occur in summer within certain wall constructions. In solid masonry walls with insulated linings orientated from east-south-east through south to west-south-west, sunshine falling on damp masonry can drive water vapour into the construction, causing condensation on the outside of the vapour control layer (Fig. 6.2). Because the vapour control layer is always supposed to be placed on the warm side of construction, any insulation behind it will become wet (BRE Digest 1994). Reverse condensation, however, has apparently not been a problem in 'no-fines' construction.

Clear night radiation condensation

At night when the sky is clear, roofs lose heat by radiation to the atmosphere (Garratt & Nowak, 1991). This can take place quickly and significantly where the roof covering is a lightweight sheeting, which is more commonly found in industrial rather than residential buildings. If it does occur it can cause condensation on the underside of the roof sheet as well as sweating on its upper surface.

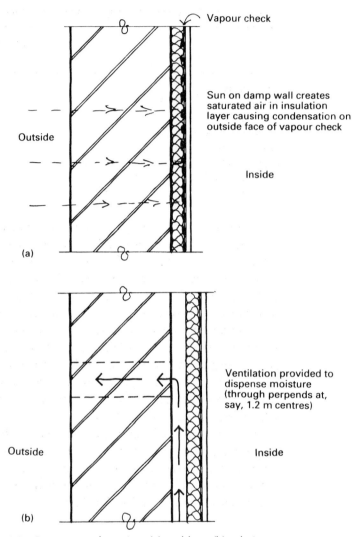

Vapour check

Sun on damp wall creates saturated air in insulation layer causing condensation on outside face of vapour check

Outside

Inside

(a)

Ventilation provided to dispense moisture (through perpends at, say, 1.2 m centres)

Outside

Inside

(b)

Fig. 6.2   Reverse condensation: (a) problem; (b) solution.

Production of water vapour

In the internal building environment, many of the normal activities of the occupants produce water vapour (Table 6.2). The use of showers (now much more common), unflued paraffin or portable gas heaters, and unvented tumble dryers causes particularly high amounts of water vapour to be produced. With the exception of unflued heaters, these activities are confined mainly to the kitchen and bathroom. It is the spread of this vapour throughout the building that is a major factor in condensation in

Table 6.2 Production of water vapour by domestic activities (BRE Digest 297).

Activity	Moisture emission per 24 hours (litres)
4 people sleeping for 8 hours	1–2
2 persons active for 16 hours	1.5–3
Cooking	2–4
Bathing/dishwashing	0.5–1
Washing clothes	0.5–1
Drying clothes (unvented tumble dryers)	3–7.5
Paraffin/portable gas heaters (2 hours' use)	1–2

many situations. This is why it is important in combating condensation that doors of kitchens as well as bathrooms should be kept closed when these rooms are being used (Garratt & Nowak, 1991).

In a new building, water vapour can also arise through the drying-out of the materials of construction. Mixing water in concrete and plaster or exposure of materials to inclement weather before and during construction can give rise to considerable quantities of water for up to a year after completion of a new building. In time, of course, the building dries out, and this source of water vapour is gradually dissipated to a suitable moisture sink. For a typical three-storey dwelling the amount of moisture generated has been estimated to be anything between 4000 and 8000 litres, and this could take over 12 months to dry out. In fact the actual figure could be much higher than this if spillages and water used for cleaning were to be taken into account (Addleson & Rice, 1991). Naturally the increasing use of dry-lining and the consequent reduction in 'wet trades' work will inevitably lower the amount of construction moisture generated in modern buildings.

Adverse effects of condensation

In temperate climates, air is generally warmer in buildings as compared with outside temperatures, and this increases the tendency of the warmer vapour-laden internal air to infiltrate throughout a building if access is available. If any of the structural elements of the building are permeable – for example, a ceiling below a roof space – the structural elements themselves, e.g. the roofs or the walls, may also become infiltrated. When vapour-laden air penetrates into such structural elements, any cooling that might occur – within external walls, for instance – would give rise to interstitial condensation (Seiffert, 1970).

The permeable materials in the building fabric will contain moisture vapour. These materials are hygroscopic, and as the vapour pressure conditions in the building change there is a constant flow of vapour in and out of such materials until a temporary equilibrium is reached.

In cold weather, external temperatures will fall and the extent of the area of the fabric with temperatures below the dew-point will increase. If interstitial condensation occurs, vapour pressure is reduced, and more vapour from inside the building may then fill the 'space' created. During warmer periods, however, moisture vapour will evaporate off the building materials back into the atmosphere of the building.

Continual wetting of building materials will lead to a number of adverse effects. The efficiency of insulation will be reduced because its thermal conductance will be increased by the moisture ingress. Clearly, the overall level of insulation has a direct effect on the probability of condensation. Thus a higher standard of insulation may prevent the temperature ever falling to dew-point.

There is also a possibility that wetting from condensation will lead to mould, decay of timber and even corrosion of metal fixings and components (see BRE Digest 369). It can also cause efflorescence to appear in unventilated 'cold' roof lofts as organic salts are leached out from the timbers.

Surface condensation is more likely on materials of high thermal capacity, such as dense concrete, when temperature changes occur. These materials take longer to heat up or cool down than the surrounding air. Therefore they frequently present cold surfaces after a cold weather spell, for example, on which condensation will occur. Conversely, lightweight permeable structures will always have temperatures closer to the ambient, but they are more susceptible to interstitial condensation because of their permeability.

Rapid corrosion of lead can occur as a result of condensation. Interstitial condensation on the underside of lead sheeting, especially in insulated roofs, can form if there is no ventilation 'layer' between the underside of the decking for the lead and the top side of the insulation (i.e. a vented 'warm roof'). Condensed water is distilled and lead will slowly dissolve in distilled water (Addleson & Rice, 1991).

Mould growth

One of the inevitable consequences of condensation is the growth of mould, and to many people this is the most obvious sign that humidity is high and condensation is occurring (BRE Digest 297). It needs to be stressed, however, that mould can also originate from other sources of dampness. For instance, dampness as a result of a leaking service pipe or seepage of rainwater can cause mould to appear on wall surfaces. But this form of mould is not usually as extensive as that associated with condensation.

Mould growth will not develop on wall areas affected by rising damp. This is because the soil salts in the rising damp moisture inhibit the growth of the mould. These surface moulds prefer uncontaminated water, and because condensed moisture is relatively pure, they form best in situations involving condensation.

The mould fungi will grow on suitable surfaces once the humidity rises above the biological limit of around 70% RH. Thus, in fact, the appearance of mould may occur at humidities well below those of water vapour saturation of the air (100% RH). Individual fungi have variable tolerances to different moisture contents, and different materials have variable equilibrium moisture contents at similar humidities. This will affect the distribution of mould growth on different surfaces in a building.

Fungus spores are present in the atmosphere at all times. The extent of such spores in a building is a function of the RH of the air and its temperature. A range of 2000–7000 spores (i.e. colony-forming units) per m^3 (cfu/m^3) in a typical mould-infected dwelling has been estimated (Lacey, 1994). However, the amount of spores in the atmosphere is highly dependent both on the prevailing cryptoclimate conditions and on previous infections, and can exceed by many thousands the range quoted.

Fungal hyphae will grow on surfaces in areas where consistently high humidities prevail, and these small strands will produce and liberate the spores, which are frequently coloured, usually black, green or an off-white. The colour of the spores and the supporting hyphae is responsible for making these mould fungi most obvious. Many species of fungi can occur, but *Penicillium* spp. and *Cladosporium* spp. are very common. *Aspergillus niger* is frequently the cause of black mould growth.

The development and pattern of mould growth is often useful confirmatory evidence that condensation is occurring in a building. The following pointers are indicative of condensation:

❑ damp or mould located in an exposed corner of a room in the shape of a crescent;
❑ the shape of mould which forms the outline of a cold bridge (e.g. lintel or column within a building structure);

❏ damp patches below window sills or on window reveals, frequently caused by the run-off of condensate from the window (Garratt & Nowak, 1991).

Normal eradication measures will be aimed at controlling the physical factors giving rise to the source of condensation. However, there is also the need sometimes for direct chemical treatment of the mould growth.

Proprietary biocidal (otherwise known as fungicidal or mouldicidal) washes (or a 5% diluted bleach solution wash) can be used on surfaces where persistent mould growth occurs (BRE Digest 370). If the decorative finishes are badly affected or damaged, they are best removed first. Fungicidal paints and wallpaper adhesives are also available, and these may help to control mould growth. The commonest mouldicides contain chemicals such as dodecylamine salicylate and sodium pentachlorphenate.

It should be emphasised, however, that there are a number of problems in using surface biocides. First of all their effectiveness varies according to the circumstances, and may be short-lived if no other condensation remedies are implemented. Second, biocides are composed of toxic chemicals, and therefore they must be used with caution and in compliance with the Control of Hazardous Substances Regulations (COSHH). Third, such treatment may cause staining due to incompatibility with iron compounds within certain substrates (BRE Digest 370).

In any event, the removal of mould growth is best done in the five stages listed by Garratt & Nowak (1991):

❏ Stage 1: ventilate the room to encourage drying and dispersal of the spores.
❏ Stage 2: remove spores from extensive growths with a vacuum cleaner.
❏ Stage 3: dampen growths with a solution of proprietary fungicide or a 1:4 solution of domestic bleach, plus a small amount of washing-up liquid.
❏ Stage 4: wipe thoroughly, apply more mouldicide and leave to dry out for a week.
❏ Stage 5: if mould reappears, it should be cleaned off with fungicidal wash at an increased concentration. Any further recurrence indicates the need for more stringent measures to cure dampness.

Care must be taken during the removal of surface growths to avoid any contamination of the air by dust and spore particles (BRE Digest 370).

Why is condensation a modern problem?

Condensation has become such a major problem nowadays essentially because of economic and technological change. Increasing energy costs have placed an economic restriction on the amount of heating that occupants can afford, and have encouraged increases in energy efficiency in buildings. Modern building standards have thus aimed at achieving higher insulation and lower natural ventilation levels. Financial pressures have also forced builders to achieve lower building unit costs since the recessions from the late 1970s. These factors have been exacerbated by changing patterns of living in homes, all of which have tended to increase the amount of moisture generated in a dwelling.

The principal influences on the emergence of condensation as a major form of dampness in buildings, therefore, are summarised as follows.

Building design and construction

The volumes of rooms in residential buildings have decreased, and the traditional heating facilities within them, such as open-hearth fireplaces, have been all but eliminated. The move away from the Parker Morris housing standards in the early 1980s has inevitably led to smaller sizes of rooms in dwellings. The minimum floor-to-ceiling height of 2.3 m is now no longer a requirement under the Building Regulations.

The forms of construction, too, have been varied. After World War II there was a major boom in the provision of non-traditional housing. This was prompted by the shortage of traditional materials and skilled labour. Non-traditional housing systems were seen as a cheap and quick alternative to conventional techniques.

Thus smaller room sizes, coupled with inadequate provision for background ventilation and poor insulation detailing, have concentrated the build-up of moisture within houses.

The growth in rehabilitation schemes has highlighted deficiencies in damp-proofing. These schemes often include modernisation works, which endeavour to make buildings more weatherproof.

Heating methods

The use of coal for heating buildings has more than halved over the last 30 years. Prior to 1950 most people carried out their domestic activities around open coal fires. Since that time, more general house heating has been achieved by using heaters fired by electricity, gas or oil. This has resulted in the reduction or elimination of open-hearth fireplaces in modern houses and their removal in many building rehabilitation schemes. It has been estimated that an open-hearth fire gives about 4.5 air changes per hour (ac/h) per room, and so their removal results in a considerable reduction in 'natural' ventilation.

With increasing energy costs, economies can easily be achieved either by switching off the heating in unused or occasionally used rooms or, alternatively, by switching heating off during the day when the house is left empty, and through the night. Energy prices have also encouraged the use of paraffin or bottle gas heaters because their initial cost is relatively low. Such heaters are more convenient to operate and transport. It has been shown, however, that portable heaters are about twice as expensive to run per hour than electric or gas fixed heaters (Garratt & Nowak, 1991).

The temperature fluctuations associated with such intermittent use lead to humidity variations, and condensation can develop during the 'colder' periods and in unheated areas (e.g. store-rooms, halls, certain corridors). The portable heaters generate much additional water vapour – 1–2 litres of water vapour – per litre of fuel consumed (Table 6.2).

Windows

It is now common practice to replace single-glazed timber or metal windows with either uPVC or hardwood double-glazed units. Double-glazing causes the inner pane in the windows to be warmer than a single pane. This means that condensation will be less likely to manifest itself on the room-side surface of the glass. All this does, of course, is shift any condensation elsewhere in the room.

Another problem with many unregulated window replacement schemes is the lack of suitable background ventilation in the frames. The absence of any such basic ventilation can exacerbate a condensation problem. All new and replaced windows should have at least 4000 mm^2 of background ventilation in the form of trickle vents in the head of each frame (see Fig. 6.7).

Draughtproofing

One of the simplest energy conservation measures is draughtproofing of windows and external doors. It is an effective and inexpensive method of reducing heat losses. It has been estimated that such measures can halve the heat losses that are due to 'natural' ventilation. However, the elimination of natural ventilation provided by unsealed windows and doors will remove another of the major pathways along which some of the vapour-laden air could escape to the outside. When installing draughtproofing it is important to ensure that humidity will not be increased in this way, and also that sufficient air is present for heating appliances (BRE Digest 306). In rooms with gas cookers or either instantaneous or storage water heaters, ventilation requirements must be followed, and draughtproofing should not be applied.

Living habits

Westerners now have much better facilities in their dwellings for washing, cleaning and cooking. These have allowed them to carry out their ablutions more regularly than they did over 20 years ago. Major improvements in the provision of basic amenities such as internal toilets and baths in houses have been made during the second half of the twentieth century. Higher living standards and smaller family sizes have allowed many householders to take a bath or shower once a day instead of once or twice a week, which was the norm at one time for many working-class people in the early 1950s (Allen, 1972). Washing appliances such as tumble dryers (which require direct venting to the outside air) are much more common nowadays. All these factors have helped to increase the amount of moisture generated in dwellings.

Expectations

As previously indicated, people are demanding more from their buildings nowadays. They are thus becoming less tolerant of deficiencies such as dampness because of increasing awareness of its known adverse health effects. What might have been accommodated 20 years ago is now no longer accepted by most people.

Admittedly, there will continue to be a small number who, because of poverty or other circumstances, are unable or unwilling to regulate the heating and ventilation in their home. Indeed, some will also continue to block internal and external fresh-air vents in the mistaken belief that this will save on heating bills. Another bad practice is the removal of overhead closers from doors to bathrooms and kitchens, and even in some cases the removal of kitchen doors themselves. Education and encouragement will be the only ways of reducing such practices.

Reducing condensation

Choices available

For all who are concerned with building construction, adaptation and maintenance, the avoidance or control of condensation and mould growth is an important objective. However, condensation is usually caused by a combination of circumstances, and so a single-factor remedy is unlikely to resolve the problem completely. Various choices are available, including ventilation, heating, structural insulation, and dehumidification.

Heating

Condensation is likely to be a problem in dwellings that use inadequate or low levels of heating (Garratt & Nowak, 1991). Thus condensation is not uncommon in housing where poverty is prevalent, because it encourages low energy usage. Improvements in the form and efficiency of the heating system will help to minimise condensation dampness.

Ventilation

Ventilation is essential in buildings for the comfort of the occupants. It can, of course, help to modify high temperatures during warmer weather periods, and fresh air is essential for certain types of heating and cooking appliances that burn gas or solid fuel. It can also assist in the removal of vapour-laden air, which might otherwise cause condensation (BRE Digest 110).

In the Building Regulations, natural ventilation via a window with at least one fresh-air inlet opening of an area of at least 5% of the floor area of a room, some part which has to be 1.75 m above floor level, is specified. Where this cannot be provided, for example in a room within the building, mechanical extract ventilation is acceptable. In a habitable room in a dwelling one air change per hour is required, but in kitchens and bathrooms three changes per hour are specified.

One ac/h ventilation can cause up to 20% of the heat loss from a traditional dwelling. In a draughty house this heat loss increases, but in a modern building without flues, ventilation may be reduced, with an improvement in effective heating.

In older houses, chimneys would act as a means of ventilation as well as a pathway for escape of vapour-laden humid air from the living rooms and kitchen. However, in many rehabilitation schemes chimneys are either removed or blocked off, and an opportunity is lost to introduce ventilation through the now-redundant chimneys.

A more useful form of mechanical ventilation is the extractor fan. These are best installed in rooms where water vapour is created: that is, kitchens and bathrooms (Fig. 6.3). Some extractors are controlled by humidistats to ensure that they are used whenever humidity levels rise above a certain level – about 65% RH. Manually controlled fans must be used during the period of vapour production and for a long enough period afterwards to be effective in removing sufficient water vapour. A 2 or 3 min delay cut-out is usually adequate.

Externally ducted extractors installed over cookers are an effective

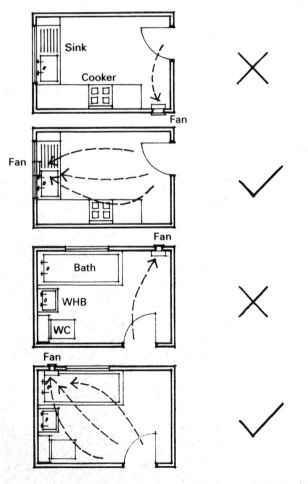

Fig. 6.3 Siting of an extractor fan in a kitchen and bathroom (GEC Xpelair Ltd).

installation, if correctly used, to reduce both humidity and kitchen smells. Kitchens and bathrooms are considered to need up to three air changes per hour to eliminate excess water vapour. However, much higher rates are needed during and after periods of cooking or washing, for example, where air changes of up to 10–15 per hour may be required. These are achievable with mechanical extract fans in such situations having a speed of at least 60 litres per second (see Table 6.9).

Internal bathrooms, toilets and kitchens must have mechanical extract ventilation (see BRE Digest 170).

Insulation

All materials transmit heat, but in varying degrees of efficiency. Solid materials transmit heat by conduction, a process in which neighbouring molecules 'pass on' temperature variations to reach a balance or equilibrium state. The part of the material nearest to the heat source will convey the heat along the material. In addition, the material will radiate heat to all colder adjacent objects, depending upon emissivity, or on to an adjacent gas which, if unrestrained, will further disperse the heat by convection.

Some materials conduct heat more rapidly than others. Usually more dense and compact materials will be good conductors. Open, porous materials contain imprisoned air bubbles, which resist the passage of heat: thus some of the best insulators are lightweight cellular fibrous materials with low structural strength.

We require the envelope of a building both to perform structurally and to act as an insulator to protect the occupants from the cold outside temperatures. Few structural components, apart from timber, have useful insulating properties, and any attempts to improve the insulation property by aeration of concrete, for instance, results in loss of strength (e.g. lightweight concrete blocks). In construction, a range of materials therefore have to be used to meet all the different requirements. In addition, air cavities are introduced to improve insulation as well as to prevent damp penetration.

The Building Regulations specify insulation requirements for exposed parts of the external fabric, allowing for windows and roof lights, to have specific U-value ratings (see the latest values listed in the final section of this chapter). The thermal transmittance or U-value of an element (a wall or a roof) is a measure of the ability of that element to conduct heat out of a building: the greater the U-value, the greater the heat loss. (Typical 'U'-values for common building construction elements are given in Table 6.5.) The position of the insulation is also important, and the relative merits of internal and external insulation are summarised in Table 6.3. A comprehensive appraisal of internal, cavity and external insulation systems is presented in BRE Good Building Guide 6.

Table 6.3 Comparison of the performance of internal and external insulations.

Internal insulation	External insulation
Warms up quickly	Slower warming
Structural component mass is kept colder	Components could become a 'thermal store' and subsequently reradiate if the component has sufficient thermal capacity
Better with intermittent heating	Reduces rapid heat fluctuation

There is no doubt that considerable benefits can be achieved in terms of both internal comfort and energy efficiency by lowering U-values in external building elements. However, this reduction of heat loss through the structure may lead to interstitial condensation unless restrictions to the flow of water vapour through the structure in the form of vapour checks and control layers are also considered.

Vapour control layers and checks

The risk of condensation occurring in or on the enclosing elements of the structure is determined by the vapour and thermal resistance of these elements which make up the structure. If insulation is to be installed in the construction, it will inevitably result in humidities increasing as long as vapour-laden air can penetrate through the structural element beyond the insulation, because it will then be at a lower temperature. One method of controlling this is to prevent, or at least restrict, the diffusion of vapour-laden air into the structure by the use of vapour control layers or checks.

There are various materials that can be used as vapour control layers, such as polythene sheeting, or aluminium foil-backed plasterboard. The vapour resistance of some building materials is shown in Table 6.4. One important principle as regards the position of vapour checks is that they should always be placed on the warm side of construction.

Table 6.4   Vapour resistance of some building materials.

Membranes	MN s/g (vapour resistance)
Aluminium foil	175–10 000
Aluminium foil in kraft paper	400–1000
Polythene sheet 250/500 g	110–450
Bitumen roofing felt	5–100
Exterior plywood	25–40
Gloss wall paint	7.5–40
Emulsion paint (2 coats)	0.2–0.6
12 mm thick fibreboard	0.4
10 mm thick plasterboard	0.35–0.5

Pitched roofs and condensation

The problem of condensation in pitched roofs can be closely associated with the installation of roof insulation, either at ceiling joist level or between the rafters (BRE Digest 180). It is further compounded if the insulation is laid down to the eaves, blocking any roof void ventilation from the eaves. It is particularly common in roofs that have sarking boards or felt under the tiles or slates. The reduction of 'natural ventilation' that would have occurred through gaps in the roof is the main cause here. The objective of roof insulation is to keep heat in the house and thereby reduce heat losses to the atmosphere through the roof. This must in the process reduce roof space temperature, and it has been shown that an average reduction of half a degree Celsius can be caused by increasing ceiling level insulation from 25 to 100 mm thickness at ceiling joist level in roofs with sarking. (The current recommendation is 150 mm thick insulation between ceiling joists.) That is why it is also important to insulate all pipework in a 'cold roof' void. Any water tanks should also be insulated on all sides except for their base, to allow heat flow to keep the water from freezing.

Over 25% of the water vapour generated in a house by the occupants reaches the roof space principally through the loft hatch, spaces around service pipes, and wire access holes to the roof void (BS 5250). Reductions in temperatures cause an increase in RH, and the risk of condensation in roof spaces is that much greater. Roof timber moisture contents would also be increased. Condensation is most likely to occur on the cold water pipes and tanks in the roof space, but mould growth or even salt deposits on roof timbers are commonly seen.

Some years ago, when condensation in pitched roofs was a relatively newly detected problem, there was a view that the best remedy was to construct ceilings with vapour barriers, which effectively prevented the vapour from reaching the roof space. However, it was soon appreciated that condensation on the surface of the vapour barrier might cause staining of plasterboard ceilings, and anyway would probably be ineffective because of 'leaks' around pipes and around the ceiling access hatch.

The more modern approach is to accept that excess water vapour must go somewhere, and that if it reaches the roof spaces it can be removed easily by ventilation. A variety of proprietary products have been produced to facilitate ventilation, both in the form of eaves ventilators (Fig. 6.4), which allow air to pass above the roof insulation at joists level, and as ridge and tile ventilators, which are essential on any lean-to or monopitched roof or roofs with parapet walls. An equivalent of 25 mm continuous openings on opposite sides of a roof are now recommended by the current Building Regulations for roofs with a pitch over 15°.

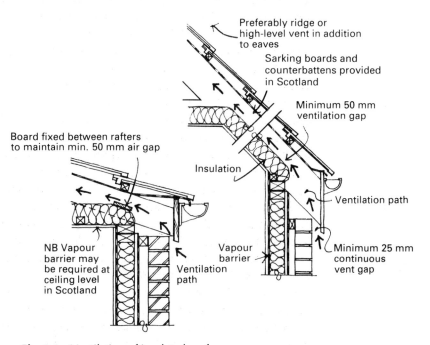

Fig. 6.4 Ventilation of insulated roof spaces.

Various external 'spray-on' systems for low-cost repairs to slate and tile pitched roofs, to prevent rainwater penetration, have been offered for many years. When battens or sarking boards are deteriorating and fixing nails are rusting away on old roofs (i.e. the roofs are 'nail sick'), the alternatives are either stripping and re-slating or a 'spray-on' resin bitumen fibreglass composition on the outside of the roof, which holds the slates or tiles in place. The latter may appear to be initially a cheaper option, but its life expectancy is uncertain, and ultimately an expensive stripping and replacing operation will probably be necessary. Such spray-on coatings could encourage internal condensation as well, by sealing up some of the natural roof space ventilation present previously via air gaps between the slates or tiles.

An alternative technique is to spray on a 5 mm thick coating of an emulsion composition incorporating a filler to the underside of the tiles. This coating holds the tiles in place and slightly improves insulation. The need to ventilate the roof space would, of course, still apply.

A recent development in new construction or major rehabilitation schemes is to use composite sarking boards that are insulated. An insulation layer between 65 and 80 mm thick (depending on the material selected) can be used, and this converts the roof space into a 'warm roof'.

Condensation and external walls

As has been discussed earlier, there can always be a risk of condensation on the surface or in the structure of a wall. This is affected by the vapour permeability of the materials of construction, and by the degree of insulation provided. The significance of any condensation that develops depends on such materials, and the major effects would be expected in a lightweight (i.e. timber frame) construction.

It has been shown that vapour checks and control layers are now incorporated in external walls to prevent interstitial condensation, and the diagrams in Fig. 6.5 illustrate the effects of these barriers. Doubts have been expressed about the effectiveness of checks of this kind on a number of grounds:

❑ that they may be installed incorrectly by builders during construction;
❑ that their effectiveness may be reduced by the penetration of wiring and piping through the barrier; and
❑ that there is evident difficulty in determining whether such barriers have been properly installed in a completed timber frame building.

However, published surveys of timber moisture contents in timber frame buildings have recorded only relatively low moisture contents, with a tendency for these to fall as the building dries out.

Investigations by the BRE are proceeding to establish the merits of constructing an inner cavity in the external walls of such buildings, in which services are fitted, to reduce the need for penetration of the vapour barriers. Another variation is to install the insulation on the outer surface of the timber frame, on the inner side of the cavity wall, forming a type of 'warm wall' construction, and trials are also in progress on this system.

In walls of brickwork, stone or concrete, the major objectives are to try to enhance the level of insulation and so improve the internal temperature, or to enable the occupants to achieve an adequate level for less use of energy. It is frequently stated that 30% of the heat losses from a building occur through the walls, and so the reduction of these heat losses should result in energy saving as well as a degree of condensation control. Various forms of in-situ insulation are possible on walls. Cavity walls can usually be insulated with one of the various forms of cavity wall insulation; solid walls can be insulated either externally or internally.

In-situ cavity wall insulation

Cavity walls in the UK were introduced shortly after the beginning of the twentieth century and became mandatory in the 1947 (English) Building Regulations for most styles. Standard 270 mm (11 in) cavity walls have a U-value of about $1.47 \, W/m^2K$, which is too high for modern energy efficiency requirements. Retrofit cavity fill insulation has been the most popular method of improving the insulation standard of such walls.

The previous Building Regulation level of $0.6 \, W/m^2K$ was achieved with cavity fill insulation (Table 6.5). Compliance with current standards would depend on the form of wall construction and the presence of double-glazing (BRE Digest 108).

The use of such cavity insulation has gone through several booms and slumps in popularity since it was first introduced in the early 1960s. The idea of filling a cavity, the original purpose of which was to provide a barrier to water penetration, was an anathema to many. Thus when reports that such insulation could cause water penetration were circulated, their caution appeared justified. However, there are many different systems of cavity fill, and the specialist contractors concerned have always maintained that, if properly and correctly applied to suitable construction, the chances of water penetration are remote – 1 in 500, according to the BRE (Pountney et al., 1988) – and are easily remedied when they do occur.

There are other potential problems with cavity fill:

❑ Gas from urea formaldehyde foam insulation may escape into the building through gaps in the walling. The fumes given off by the gas can be very irritating for some occupants.

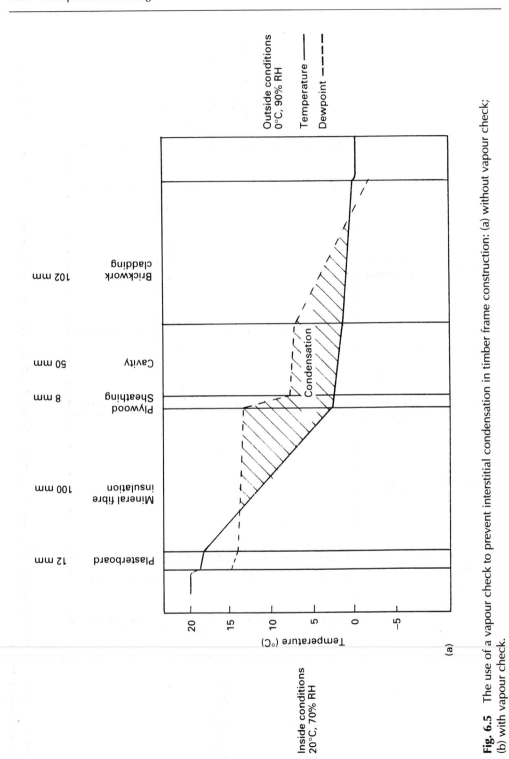

Fig. 6.5 The use of a vapour check to prevent interstitial condensation in timber frame construction: (a) without vapour check; (b) with vapour check.

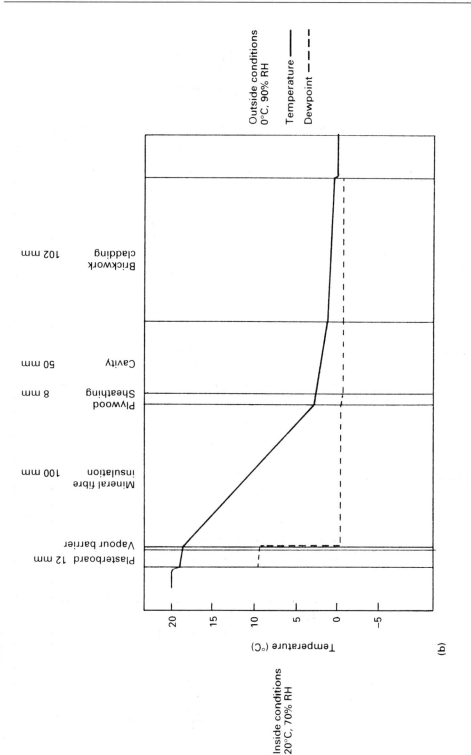

Fig. 6.5 (cont'd)   The use of a vapour check to prevent interstitial condensation in timber frame construction: (a) without vapour check; (b) with vapour check.

Table 6.5   Typical U-values for some building construction elements.

	U value[a] (W/m^{2}K)
Solid walls	
220 mm brickwork	
with 13 mm interior dense plaster	2.14
and with tile hanging	1.70
and with 13 mm dense external render	2.02
Cavity walls	
270 mm cavity brickwork	
with 50 mm cavity	
and 13 mm interior dense plaster	1.47
105 mm outer brickwork	
with 50 mm cavity	
and 100 mm inner concrete blockwork (600 kg/m^3)	
and 13 mm interior dense plaster	0.96
(with 13 mm lightweight plaster	0.93)
(with 25 mm polystyrene insulation	0.83)
105 mm outer brickwork with 50 mm mineral wool insulation in cavity	
and 100 mm aerated concrete blocks 600 kg/m^3	
and 13 mm lightweight plaster	0.47
105 mm outer brickwork	
with 50 mm cavity	
and bituminised breather paper barrier	
and timber frame with 100 mm mineral wool insulation	
and 13 mm plasterboard	0.34
Pitched roofs	
Tiles on battens, roofing felt and rafters	
and 13 mm plasterboard	1.5
with 100 mm glass fibre insulation	0.35
Flat roof	
Bitumen felt, timber deck	
150 mm joists	
and 13 mm plasterboard	1.5

[a] The lower the U-value figure, the better is the standard of insulation.

❏ External filler holes are slightly disfiguring, and if not properly sealed may encourage rainwater penetration.
❏ Mortar droppings and other debris in the cavity prevent a complete fill of the void, which will create cold bridges.

Factors that may be conducive to penetrating damp in such walls are more significant in exposed positions, where driving-rain conditions saturate the outer skin of brickwork and could cause water leaks through and into the cavity. High permeability of the mortar and brickwork, incorrect positioning of wall ties and cavity trays, and the presence of mortar bridging may allow water penetration to become much more significant when cavities are filled with insulation.

In England and Wales, the filling of wall cavities in existing buildings is permitted under a 'type relaxation directive' for conventional buildings less than three storeys in height. Several systems are available, and these are compared at length by Cairns (1992). In summary, the main types of cavity fill insulation are:

❑ urea formaldehyde foam;
❑ mineral wool granular fill;
❑ polystyrene beads;
❑ polyurethane foam.

Urea formaldehyde foam was one of the most widely used initially. It is made up of a water-based resin with a hardener added immediately prior to injection through a series of holes drilled in the external leaf of the brickwork. After application, it hardens rapidly and dries. Sometimes drying may lead to shrinkage and possibly cracking, perhaps allowing water penetration across the cavity to occur. This system should be used only in cavity walls of brick and block of buildings up to three storeys and of moderate exposure (BS 5618). In addition, it should not be used to fill the cavity of any building within 7 km of the sea to minimise severe exposure, particularly where the outer leaf is not rendered. The inner skin should be constructed in such a way that there are no gaps through which formaldehyde vapour produced during the application process could infiltrate into internal and occupied spaces.

Mineral wool or glass fibre pellets have proved to be a popular alternative to foam. The pellets can be blown into the cavity through 12–15 mm diameter holes in the outer brickwork.

The other main granular fill option is to use expanded polystyrene beads blown into wall cavities. These beads are free flowing, and so fewer filling holes are needed, although beads may be 'lost' through unseen holes and gaps in the inner leaf of the masonry. A binding agent is used in some systems to fix the beads in situ, but voids then develop in the cavity fill. Polystyrene granules are irregularly shaped, so they do not flow well, and are less likely to 'escape'.

Polyurethane foam has been used both for insulation and to stabilise walls where wall ties have corroded (see BBA certificate 85/1567). It foams in situ and fills the cavity, adhering strongly to the masonry. (See also Chapter 5.)

Tests carried out by the BRE suggest that rainwater penetration is more likely with urea formaldehyde foam than with any other type of cavity fill. But a survey of 40 000 UF-filled low-rise dwellings reported only a 0.2% incidence of water penetration where cavity fill had been installed in situ to existing occupied buildings.

Factors affecting water penetration include the degree of exposure, structural faults, frost damage, general condition of brickwork and pointing, gaps around window frames, pipes, vents and flues, and the continuity of these defects into the inner leaf of the wall. Obviously these defects should suggest the unsuitability of a building for cavity insulation.

If rainwater penetration occurs as a direct result of one or other of these faults in a cavity-filled property, the first remedial action must be to determine the presence and extent of the defect, and take appropriate measures. Where water penetration has occurred through walls otherwise apparently suitable and free from visible defects, successful remedies involve the re-injection of the cavity insulation, where there are voids in the insulation, for example; clearing out of cavities where the mortar bridging is responsible; or coating the wall with a silicone water-repellent. Only very rarely is overcladding, new render or the removal of the insulation necessary as a remedial measure.

A particular problem with in-situ cavity wall insulation has arisen in Scotland, where some timber frame houses have been insulated. While under current codes of practice and British Standards such in-situ insulation installation are not recommended, some older timber frame houses have been insulated because the standards of the original insulation were much lower than would be used today. With such construction, any penetrating moisture might cause decay in the timber frame, which was often constructed in those days from untreated softwood.

The mere fact of the presence of this cavity insulation in such houses has made surveyors and building societies reluctant to approve mortgage advances to prospective buyers, and such properties have been difficult, if not impossible, to sell. There has been a demand to have the insulation removed in order to overcome the problems, but this would be very disruptive, time consuming and thus expensive to do. It would probably involve inevitably the removal also of the outer cladding leaf of brickwork. The detection of dampness problems in such cases can be facilitated by thermographic cameras, as described in Chapter 9.

In all types of construction, the presence of cavity wall insulation would make the outer skin of brickwork or masonry colder, and probably higher moisture contents would prevail in this element. This could accelerate frost damage of brickwork, especially in areas where water penetration might occur and where drying out might be restricted by a dense render or an impervious paint. In such situations, the outer leaf must be protected as far as possible from penetrating dampness. This could be done by coating the external wall with a silicone sealer or a two-coat render system.

Insulation of walls in new construction

Nowadays there is an increasing emphasis on improving the insulation standards in new construction. As a result of this influence, the design and construction of external walls has had to change. The current Building Regulations now require external walls to have a U-value of $0.35 \, W/m^2K$, which is a significant improvement on the $0.6 \, W/m^2K$ rating for walls in the late 1980s (Table 6.6).

Table 6.6 Comparison of wall insulation systems.

	U-value (W/m²K)
220 mm solid wall with internal plaster	2.14
270 mm cavity wall with internal plaster	1.47
with cavity fill	
with 50 mm UF foam	0.56
with 50 mm blown mineral fibre	0.53
with 50 mm EPS beads	0.51
with internal wall insulation	
27 mm thermal board	
fixed by adhesive	0.54
fixed by battens	0.52
with external insulation	
30–65 mm thick	0.55–0.35
current requirements in Building Regulations	0.35

One of the great advantages of timber frame construction has been the potential to meet these new standards and even exceed them. Conventional brick and block construction has met the new requirements in several ways.

The use of lightweight aggregate and aerated blocks is widespread. However, as insulation properties increase, the density and strength properties of such materials decrease. Wider blocks and blocks incorporating insulation, either in internal cavities or as composite bonded aggregate blocks, have been developed. One perceived problem of using such blocks is the risk of pattern staining caused by the much more heat-conductive mortar, as well as cold bridging at lintels and other junctions of components.

Wider cavities, up to 80 mm wide, providing space for semi-rigid partial-fill cavity insulation, have also been employed. Semi-rigid mineral fibre batts, cellular glass slabs, glass fibre batts or expanded polystyrene boards are available as well. A 50 mm wide clean cavity is essential in wall cavities that have such cavity insulation fixed to the inner leaf of the masonry. This is required to allow for the dispersion of any vapour or condensation by ventilation in the cavity, as well as to prevent direct water penetration. Fire stopping in the cavity of some cases (e.g. timber frame buildings and overcladding schemes) will be required, usually at every storey and at about 8 m centres horizontally.

Another method of improving insulation of walls in new construction is by in-situ cavity wall insulation. However, this relatively recent practice may delay drying out for a few weeks. The BRE experience is of a higher percentage of cases of damp penetration (c. 3%) in such recently constructed and insulated walls (Pountney et al., 1988). This is probably a reflection of the poor standard of some modern bricklaying, which gives rise to defects such as bridged cavities, misplaced cavity trays and inadequate pointing.

Insulated dry-lining

Dry-lining is an old-established method of improving insulation of walls. In Scotland and the North of England it is the traditional method of internal finish for the thick stone walls of many Georgian, Victorian and Edwardian buildings. It has some merit of being a suitable method for in-situ improvement of insulation in many housing rehabilitation schemes, as well as being useful for achieving adequate U-values in new construction. The disadvantages with this method are that it reduces the living space in a room; it can be very disruptive to install in an occupied dwelling; and it does not keep the main fabric of the building at a stable temperature level.

Insulated dry linings can provide much better levels of thermal efficiency, and their resistivity permits rapid build-up of heat in a room. They can also be useful in dealing with cold bridges (see below). However, to avoid interstitial condensation in such warm-side insulated linings, a vapour check will be needed, and provision must be made for ventilation of the insulated cavity.

Two types of lining system are used. The traditional method was to use plasterboard, either foil-backed or with a vapour check, fixed to treated timber battens. Prior to fixing the plasterboard, insulation can be fitted in the formed cavity between the battens. A more modern approach is to use a composite sheet of plasterboard and insulant, such as expanded polyurethane, which can be fixed directly to the wall, either mechanically or with a plaster dabs adhesive, or onto timber or metal battens. Gyproc Fireline Board 35 mm thick incorporating a 25 mm thick insulation board, having relatively low toxicity in fire, and fixed with plaster dabs, is a typical modern dry-lining option.

A particular problem exists when dry-lining a wall that has become damp, especially in an earth-retaining situation in a basement. A form of tanking is then necessary to protect the lining from penetrating dampness, and a ventilated cavity is required between the tanking and the insulation.

Cold bridges

Cold bridges (or thermal breaks) occur at parts of the fabric that allow heat to escape from the fabric at a much higher rate than elsewhere. They are commonly found at openings where boot lintels have been used, at junctions around windows and doors and external wall areas (BRE, 1994). If left unattended, they will eventually give rise to local condensation pattern staining and mould growth. Some examples of cold bridging are shown in Fig. 6.6.

External insulation of walls

Cavity wall insulation appears to be the most cost-effective method of improving the thermal performance of walling (BRE Good Building Guide 5). However, some of the housing stock in the UK still has solid external walls in either brickwork or stone. External insulation is another method proposed for such construction, and reductions in U-values from around 2.2 to around 0.35 W/m^2K are possible.

External insulation must be weather resistant; indeed this is one of the advantages of the system when the original structure may have already deteriorated to the stage where essential maintenance and repair is required to cure dampness.

Several systems are in use. Most involve the application of a slab to the

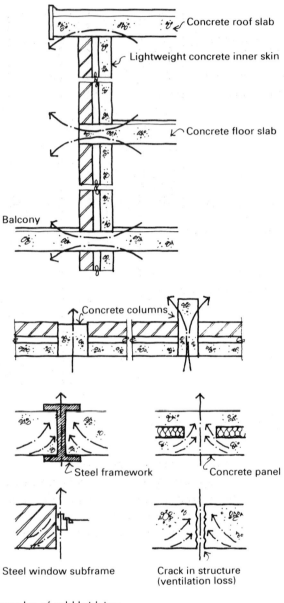

Fig. 6.6 Examples of cold bridging.

external wall of the building. These slabs are commonly expanded polystyrene, but polyurethane, foamed glass polyisocyanurate or mineral fibre are also used (Fig. 6.7). These slabs are usually fixed mechanically to the wall; edges are protected with galvanised or, better still, stainless steel strips. A glass fibre stainless steel mesh, lath or scrim

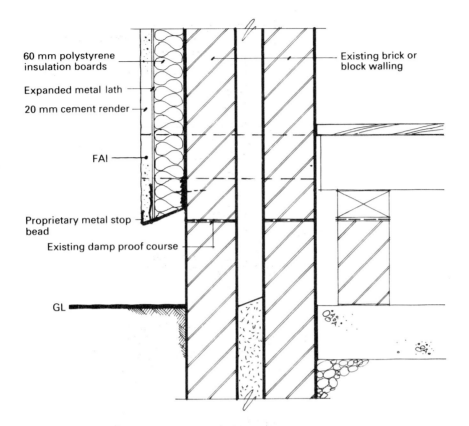

60 mm polystyrene insulation boards

Expanded metal lath

20 mm cement render

FAI

Proprietary metal stop bead

Existing damp proof course

GL

Existing brick or block walling

Fig. 6.7   Externally applied wall insulation.

provides a reinforcement for the rendering undercoat. This is of course an expensive system of insulation, and can probably be justified only if the decorative and weatherproofing features are needed as well. The general insulation performance characteristics of the various systems discussed in this section are to be found in Fig. 6.7.

Glazing

Perhaps the simplest form of 'dehumidification' is a single-glazed window. It is one of the most effective tell-tales for condensation. For this reason it is best to use single-glazing rather than double-glazing in bathrooms and kitchens (Garratt & Nowak, 1991).

In cold weather conditions, the glass represents the coldest surface (U-value 5.6 W/m^{2}K) on the inside of the house, and will therefore attract considerable condensation. Various proprietary systems have been developed to dispose of this condensed water, either through a special subsill or through a siphon system that discharges directly through the glazing. Discharge of such water through special holes drilled in a timber sill is not recommended, because it might cause decay to develop.

Generally, it is the heat losses through the windows and doors that are the overriding consideration when attempting to improve insulation and heating standards. It has been calculated that heat losses amounting to 20% occur through windows, and to this can be added losses as a result of draughts around ill-fitting window frames. The materials used for the frames are also significant. Metal frames will act as a further avenue for heat losses; uPVC and wood have better performance in terms of heat loss resistance.

Secondary double-glazing is probably always cost-effective, because it is usually cheaper to install, and the large air gap between the glazing achieves greater sound as well as thermal insulation. However, as the 'gap' is increased, convection movement of the air can occur, so that extra thermal insulation is not achieved to the same extent. It should be appreciated that double-glazing can only reduce by about one third the total heat losses through windows.

Dehumidifiers

The use of dehumidifiers for controlling condensation has become more prevalent in recent years. They can improve conditions by reducing the relative humidity of the air in a room. In a dehumidifier, humid air is blown or drawn by a fan over a cold coil, which causes condensation on its surface. This condensed water is then collected and disposed of in two main ways: (a) it can be collected in a deep tray which is emptied regularly; or (b) it can be vented outside in 'plumbed-in' dehumidifiers.

There are, however, a number of problems with using dehumidifiers, and for these reasons they should be seen as a short-term measure:

❑ Their performance is critically dependent on the cryptoclimate conditions in the room (Garratt & Nowak, 1991). Dehumidifiers can perform adequately in a well-heated room with high relative humidity, but they are much less effective in a poorly heated room.
❑ They are costly both to purchase and to run.
❑ They can be noisy.
❑ They can take up valuable floor space.
❑ Their reservoirs need to be emptied regularly.

Ground floors

The insulation of ground floors is best achieved at the time of construction. Solid ground-supported in-situ concrete slabs can be best constructed incorporating a resilient underslab of expanded or extruded polystyrene boards as insulation. This also provides a thermal barrier beneath pipes or cables. Vertical perimeter insulation (25 mm thick) around the slab and screed will be necessary to cope with thermal and moisture movements in the floor, as well as to reduce the heat losses that are always greatest through the outside edge of the floor adjacent to the external walls.

Overslab insulation consisting of extruded (not expanded) polystyrene boards combining a vapour check, and with an overlay of chipboard or high-quality hardwood tongued-and-grooved boarding, laid above a damp-proof membrane in the slab, provides a quick thermal response, which is especially useful with intermittent heating. Such a system can be used as a floating floor in existing buildings, provided detailing problems at the doorways, skirtings and stairs can be overcome.

In solid ground-sported floor slabs containing underfloor heating, it is imperative that the concrete is allowed to dry out properly prior to the installation of any screed or timber floor covering. The concrete or screed needs to have an RH value well below 75% before these finishes can be installed in such a floor, otherwise there is a risk of interstitial condensation occurring on the underside of the boarding or within the screed (see Appendix L).

Suspended timber floors constructed with either square-edged or tongued-and-grooved boarding often have poor insulation properties because of the gaps that develop as a result of shrinkage of the timber boards. The use of overlays of hardboard, carpet and carpet underlay or other floor finish will improve the insulation properties of the floor. Techniques have also been developed for installing insulation on the undersurface of suspended timber ground floor boards, but adequate access to and space within the subfloor void would be necessary for the installation.

The habit of closing fresh-air vents of suspended ground floors during winter months, with a view to improving internal heating by possibly reducing draughts through the subfloor void, is very undesirable. This practice inevitably increases the risk of condensation and decay of organic materials in the subfloor void, especially if the air vents are not unblocked during the warmer periods of the year.

Selection of remedial measures

From the foregoing it can be seen that there are a variety of measures that could be chosen to deal with any particular condensation problem. Some of the options that exist to improve the condensation situation in existing buildings are outlined in Table 6.7.

Table 6.7 Typical condensation remedial measures.

Measure	Method	Special notes and problems

Ventilation:
  living room (1 ac/h)
  bathroom/kitchen (3 ac/h)
	Humidistat-controlled extract fans (delayed cut-off)
	'Built-in' extraction ducts (use redundant chimneys)
	Roof void ventilation to BS 5620
	Direct air circulation from roof void
		Draughtproofing may interfere
		Mechanical extract ventilator requires maintenance

Insulation: higher standard	Overcladding Cavity wall insulation Dry lining Double-glazing Ceiling and roof insulation	Cold bridging metal windows without thermal breaks
Reduce vapour directly	Dehumidifier Single-glazing 'drainage' Closers on kitchens and bathroom doors	Avoid using portable gas and paraffin heaters and unvented tumble dryers
Heating: higher standard	Electrical panel heaters Vented gas systems Central heating	

Many factors, concerned both with the building and with its occupants, will affect the choice. Furthermore, one measure taken in isolation might be totally ineffective. For example, recent studies by the BRE have shown that while increased ventilation may remove water vapour-laden air, it leads to lower internal temperatures and possible discomfort for the occupants (Garratt & Nowak, 1991).

At lower levels of heating and insulation, as frequently found in houses constructed to pre-1973 Building Regulations standards, increasing ventilation will not prevent condensation and mould growth because of the lower temperature levels that prevail. Better heating and insulation are necessary.

However, at higher standards of heating and insulation some ventilation is essential to prevent condensation, and the greatest benefit is obtained at the rate of one air change per hour. This is illustrated in Fig. 6.8 (Garratt & Nowak, 1991).

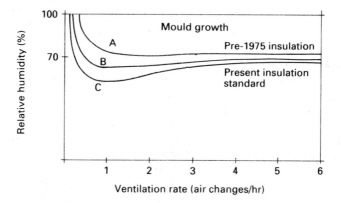

Fig. 6.8   RH related to air changes (Garratt & Nowak, 1991): A, unheated bedrooms in flats and bungalows; B, unheated bedrooms in houses; C, minimally heated bedrooms in flats and bungalows.

Whatever systems are finally selected, for the control of condensation in buildings to be effective they must be understood and used intelligently by the occupier. Too often, in tenanted properties, the unsatisfactory living conditions are exploited for what they are, and little attempt is made to reduce or control them. At the same time, there are several simple and relatively inexpensive remedies available that, with goodwill, can prove very effective. The new building regulations and codes are producing buildings now that should not suffer from condensation to the same extent in the future. The main problem to be faced is how to improve existing stock to standards appropriate to the 1990s and beyond, and at the same time to educate property users in condensation avoidance.

Relevant changes to the Building Regulations

Background

Since the mid-1980s there have been a series of major changes to the various British building regulations. For the sake of convenience, the Building Regulations 1991, applicable to England and Wales, are being cited here. The improvements mentioned below are also applicable to the Building Standards (Scotland) Regulations 1990.

There are three significant areas of change that are of relevance here:

❑ SAP energy ratings;
❑ new ventilation requirements;
❑ new U-values.

Standard Assessment Procedure (SAP)

All new dwellings now require an energy rating calculated under the Government's SAP before obtaining a building warrant. SAP is a rating that provides a convenient indication of how energy efficient a building is, on a scale of 1–100: the nearer the number is to 100, the more energy efficient is the property. An SAP rating above 80, for example, would be considered an acceptable target; one below 60 would indicate that thermal improvements are required.

SAP takes into account fabric heat losses, ventilation rates, solar and internal gains, along with heating requirements. These are then considered as an overall thermal efficiency package for the building. The thermal standards of individual elements (walls, roofs, windows, etc.) can be modified to offset one another or combined to enhance the resultant energy performance of the dwelling. The assessment procedure rating is based on the total running costs per square metre of floor area excluding lighting and appliances for a fixed location within the British Isles. The Owens Corning software program indicated earlier includes an SAP calculation facility.

New U-value targets

The new target U-values for dwellings are dependent on the SAP rating achieved. This method gives the designer/builder a degree of flexibility for the proposed construction. Table 6.8 illustrates the effects on U-values where different SAP ratings are used. It will be seen therefore that higher U-values (i.e. ones relatively less thermally efficient) can be allowed if the overall SAP rating is high (i.e. the dwelling is relatively more thermally efficient as a whole).

New ventilation requirements for dwellings

The consequences of improving the thermal efficiency of dwellings are fairly obvious. Buildings will become more airtight and less prone to

Table 6.8   New U-values for typical building elements.

Element	U-values for dwellings	
	SAP rating over 60	SAP rating under 60
Roof lofts	0.25	0.2
Skylights	3.3	3.0
Pitched roof slopes	0.35	0.2
Flat roofs	0.35	0.2
Walls	0.45	0.45
Soffits	0.45	0.35
Floors	0.45	0.35
Garage doors	3.3	3.0

leakage through the fabric as seals are required around openings. This increases the risk of condensation occurring.

In order to prevent problems with condensation, Part F of the English Building Regulations has been amended, and in summary the changes are as follows:

❏ Further clarification and additional requirements are given for background ventilation.
❏ Passive ventilation can now be used as a credible alternative to mechanical extraction.
❏ Further guidance is given on reducing the risk of flue gas spillage from open-flued appliances used in conjunction with mechanical extraction.
❏ Requirements to ventilate utility rooms (entered from within a dwelling) are now excluded.
❏ Provision for opening windows in kitchens has been added as a supplement to extract ventilation.

Table 6.9 shows the new requirements.

Conclusions

So long as energy efficiency measures and indoor washing practices continue, condensation will be an ongoing problem in buildings. If anything, the intensity of these influences will probably increase. For these reasons condensation is likely to remain the most severe form of dampness to affect buildings for many years to come (Melville & Gordon, 1973; RICS Building Surveyors Division, 1979).

Table 6.9   New ventilation requirements.

Room	Rapid ventilation (e.g. openable windows/doors)	Background ventilation (mm^2)[a]	Extract rates: fans or passive stacks
Habitable room	1/20th floor area	8000	[b]
Kitchen	Opening window (no minimum size)	4000	30 l/s adjacent to hob 60 l/s elsewhere or PSV[c]
Utility room (access via dwelling)	Opening window (no minimum size)	4000	30 l/s or PSV
Bath/shower room	Opening window (no minimum size)	4000	15 l/s or PSV
Sanitary accommodation (separate from bathroom)	1/20th floor area or 6 l/s mechanical extract	4000	—

[a] Alternatively provide a minimum of 4000 mm^2 to all rooms in table with an overall average of 6000 mm^2 per room for the dwelling.
[b] For internal rooms (i.e. those rooms not having any windows to the outside), provide either 15 min overrun to the mechanical extraction indicated in the table, or PSV, or an open-flued heating appliance. In all cases some form of air inlet is required.
[c] PSV = passive stack ventilation.

Although the range of causes of condensation is very wide, it is on the whole 'a combination of occupiers' activities and building design which create the conditions which lead to condensation and mould' (Garratt & Nowak, 1991).

One effective way of understanding the incidence and characteristics of condensation outbreaks is to consider typical examples. Twelve such cases are presented in BRE's excellent report on condensation (Garratt & Nowak, 1991). Building professionals who have to investigate condenstion problems would be well advised to consult those examples and use the handy checklist pro-formas that are available from the BRE. In addition, the advice on risk factors and defects avoidance given by the BRE (1994) and HAPM Ltd (1991) should be followed when undertaking new-build works.

Chapter 7
Rising Damp

Background

In the eyes of the public, rising damp could challenge condensation as being the most notorious dampness problem to affect dwellings in the UK. This is partly due to the efforts of specialist contractors promoting their particular system for the control of rising damp. Also, a greater awareness by property owners of the presence and adverse effects of dampness in buildings has contributed to this situation. Such publicity has highlighted a problem that has existed ever since the first wall was built. Only in recent times, though, when our standards of what are perceived to be 'adequate living conditions' have risen, has much importance been attached to the elimination of rising damp in buildings.

The public health legislation in the late nineteenth century was the first attempt to make the insertion of a physical damp-proof course in new housing construction a statutory requirement. Despite this, however, the presence of a damp-proof course was not always assured in all buildings. Many of the original damp-proof courses, being of poured or sheet bitumen, have become brittle with age. As a result, and because settlement of the building has occurred, many have failed. Moreover, numerous existing and potentially effective damp-proof courses have been bypassed by adjacent building alterations, or by changes in ground levels. These are just some of the reasons why rising dampness is still a problem today.

Despite these influences, a few words of caution need to be stated at this point. Notwithstanding the reasons for the apparent continuing prevalence of rising dampness, there is a growing body of opinion supporting the view that rising damp is not the most significant form of dampness in British buildings (Howell, 1995). According to the BRE, it accounts for only about 10% of dampness problems investigated, and very few of those are still active (Rowland, 1996). Furthermore, some researchers have indicated that some retrofit chemical dpc installations may not have been fully justified (Rickards, 1989; Houston, 1990).

This is not to imply that rising damp does not exist. Rather, impartial studies in recent years have indicated that it is not as severe a problem as perhaps the public and some building professionals (particularly those carrying out mortgage valuation surveys) might generally perceive. In comparison with condensation and rainwater penetration, rising damp is fairly rare (Howell, 1995).

Therefore it is probable that the reasons why rising damp is not as pervasive as these other forms of dampness are as follows:

❑ The majority of buildings in the UK have some form of original dpc. Even bridging or lack of continuity between dpc's/dpm's would cause only localised rather than widespread incidences of rising damp in a building.
❑ Failures in these dpc's would need to be severe and extensive to cause major and general manifestations of rising damp in a wall. There is no evidence that suggests that such failures are occurring on a large scale.
❑ The problem of rising damp in walls caused by defective or missing dpc's can be combated by reducing the subsoil moisture content.

Granted, it is a natural phenomenon that moisture present in the ground will rise through any porous material in contact with the subsoil. The height of such moisture in the walling is related to the forces of capillarity. Gratwick (1966) has shown that rising dampness is caused by the presence of electrical charges, and he has suggested that they are associated with the surface-tension forces that induce capillarity.

If a small glass tube or capillary is inserted in water, the liquid will rise up the container as a result of the forces of surface tension, which affect the angle that the water surface makes with the capillary wall (Fig. 7.1). Generally, the narrower the tube, the higher will be the rise against the forces of gravity. The height that the water achieves is determined by the

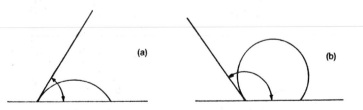

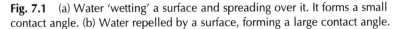

Fig. 7.1   (a) Water 'wetting' a surface and spreading over it. It forms a small contact angle. (b) Water repelled by a surface, forming a large contact angle.

hydrostatic pressure of the water already in the container counter-balancing the forces pushing it upwards – which, in fact, is twice the surface tension of the water (Figs 7.2, 7.3).

The capillaries that are present in the walling arise partly from the porous nature of the masonry units – stone or brick, for instance – and partly from the mortar. Moisture movements that are associated with drying out, shrinkage and thermal movements in response to weather and climatic changes all play their part in producing more cracks.

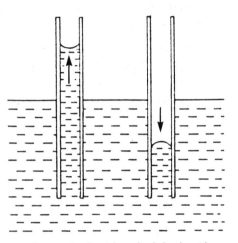

Fig. 7.2   Capillarity is forcing the liquid up the left tube. The repellent surface of the tube on the right is rejecting the fluid.

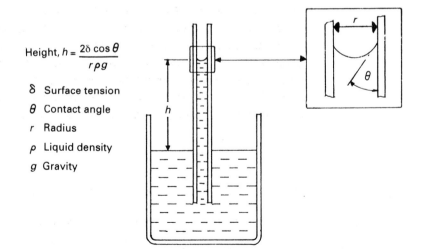

Height, $h = \dfrac{2\delta \cos \theta}{r\rho g}$

δ Surface tension

θ Contact angle

r Radius

ρ Liquid density

g Gravity

Fig. 7.3   The formula for calculating the height to which a liquid will rise up a capillary.

The measured pore size for bricks and mortar is in the range 0.001–0.01 mm. It is believed that it is the shrinkage cracks that provide the major pathways for rising damp, and the average size for these has been estimated as 0.01 mm. An approximation based on the formula in Fig. 7.3 can be used to estimate the expected height of the dampness. Thus for a pore of radius 0.01 mm, a height of 1.5 m would be expected. For a pore size of 0.001 mm, 15 m would be the theoretical height. However, micropores will have a greater resistance to flow of moisture, so a smaller quantity would rise. In practice, the water vapour evaporates off the wall, and this reduces the actual height reached, particularly on thinner walls where evaporative surfaces are larger in relation to the total volume of the wall. That is why 1200 mm is the maximum height that would normally be expected for rising damp.

The height of the rising damp will also be influenced both by the amount of moisture in the soil and by the height of the water table (Table 7.1). These subsoil conditions can vary depending on the time of year and the recent weather conditions. In addition, any leaks in the below-ground drainage and water supply services will increase the moisture content of the subsoil.

Table 7.1   Factors affecting the height to which dampness may rise in a wall.

Dampness rises further	Dampness height reduced
Higher soil moisture content > 20%	More heating and ventilation
Higher soil and ground levels above dpc	Lower water table
Timber dadoes and dry lining especially with polythene on internal walls	Thinner walls
Dense rendering on the wall	Efficient land drainage around the building
Impervious wallpapers and paints	
Higher ambient humidity	
Pore-blocking salts	

The minimum subsoil moisture content percentage required to induce rising damp has yet to be established conclusively. The indications are that it is probably in excess of 20%. Research carried out in the USA as far back as the late 1940s, for example, found that over 20% subsoil moisture content was required to cause excessive humidity levels in uncovered crawl space surfaces (Rose, 1994). Clearly this is an area requiring much more detailed investigation by independent authorities.

Another influence on the degree of rising dampness is the type and quality of subsoil. Sandy subsoils tend to be very absorbent and, when wet, will provide an ideal moisture reservoir/sink under and around the substructure of a building. This can facilitate damp to rise in the wall if it does not have an adequate dpc. Preliminary investigations suggest that moisture does not permeate as effectively through cohesive soils.

Naturally, much more research is required to confirm or confute these hypotheses. To this end, empirical studies into rising damp diagnosis and chemical treatments are being undertaken at South Bank University, London, through a research grant from the Department of the Environment. The BRE, of course, has ongoing investigations in this area.

Climatic factors internally as well as externally can affect the rate of evaporation off the wall, and if high humidities prevail, the dampness may rise further. Any obstruction to the evaporation, by a dense render or matchboarding dado, for instance, would tend to drive the water further up the wall. Conversely, if heating levels are increased this will assist evaporation and reduce the height to which dampness will rise (Table 7.1).

Moisture in the soil contains ions of dissolved mineral salts. In fact there is a wide range of salts naturally present, which are derived from the type and geological origin of the soil. In most soils, the salts that are present are combinations of sodium, potassium, calcium, phosphates, chlorides, nitrates, sulphates and carbonates.

The most characteristic salts from the soil found in rising damp are nitrates and, to a lesser extent, chlorides. Nitrates of course can emanate from fertiliser but are also a natural constituent of the soil itself. They are unlikely to be found in masonry unless they have originated in the ground. The only exceptions arise if soil has accumulated in roof valley gutters, for example, or where defective soil pipes from WCs have allowed nitrate-containing water to penetrate the masonry. Chlorides are found naturally in masonry to a very small extent, so their presence can also be regarded as characterising rising damp. The presence of chlorides, however, is less definitive than that of nitrates.

The salts have a significant effect near surfaces where rising damp moisture is evaporating. In these areas, the salts accumulate and eventually may block the pores and capillaries through which the water is evaporating. This pore-blocking effect may cause more dampness to rise even further up the wall. Calcium sulphate is the most likely candidate salt for this effect, because it is only sparingly soluble in water, and comes out of solution first.

Both nitrate and chloride salts of calcium, potassium and sodium are much more soluble than the comparable sulphate salts, so these are more likely to remain in solution longer in the rising damp moisture. In fact, because they are less likely to crystallise out on drying, it is the less soluble salts, especially sulphates, that form the characteristic efflorescence that we associate with dampness in brickwork (Table 7.2).

Table 7.2   Variations in the solubilities in water of salts commonly found in damp masonry.

Salts	Calcium	Potassium	Sodium
Sulphates	Low	Low	Medium
Nitrates	Medium	Medium	Medium
Chlorides	Medium	Medium	Medium
Carbonates	Very low	High	Medium

✗ Another and perhaps even more significant effect of these salts is as a result of their hygroscopic action. They have the property of absorbing water vapour from the atmosphere, and the humidities at which this hygroscopicity will commence at 20 °C are shown in Table 7.3. It can be seen that chlorides and nitrates are particularly hygroscopic. This means that wherever these salts accumulate they add considerably to the moisture level in the wall.

Table 7.3   Humidities at which hygroscopicity of salts that commonly occur in masonry commence (at 20 °C).

Salts	Chemical formula	RH (%) at 20 °C
Potassium sulphate	K_2SO_4	97
Sodium sulphate	Na_2SO_4	93
Sodium chloride	$NaCl$	76
Magnesium nitrate	$Mg(NO_3)_2$	55
Magnesium chloride	$MgCl_2$	33
Calcium chloride	$CaCl_2$	32

Note: Magnesium and sulphates most probably originate from the masonry, whereas nitrates and chlorides derive from ground moisture.

As these salts accumulate at places where drying-out of rising damp is occurring, their hygroscopic effects are most evident on drying surfaces and at the apex of rising damp in the wall. This has been illustrated in BRE Digest 245. Typical moisture profiles in which the extent to which dampness in a rising damp situation can be attributed to either the hygroscopic moisture or the capillary moisture are shown in Fig. 7.4. From these diagrams it can be seen that at the base of a wall where rising

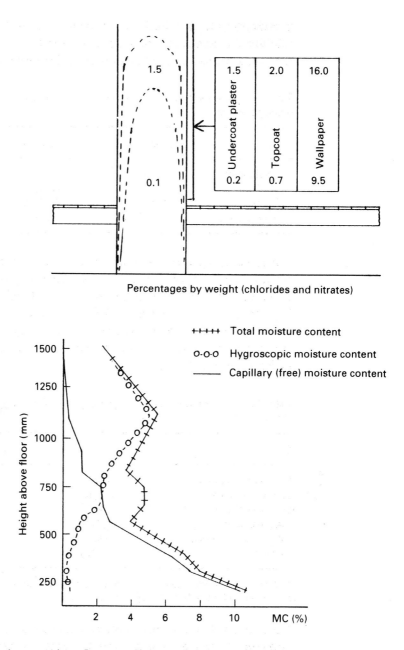

Fig. 7.4 The influences of hygroscopic salts on the total moisture present in a wall subject to rising dampness (after BRE Digest 245; and Kyte, 1984).

damp is present, the major factor in the dampness is the actual capillary moisture. Further up the wall, the influences of the hygroscopic moisture increases, and it forms the major component. Salt accumulations are seen to be at the peak of the rising damp, and on the wall and wallpaper where the greatest evaporation has taken place.

It is possible that the greater accumulations of sulphates on drying-out of sulphates of masonry is in part due to the fact that the more hygroscopic salts such as sodium and potassium chlorides may redissolve as a result of the accumulations of hygroscopic moisture under conditions of high humidity, and diffuse back into the walls, leaving the less soluble sulphates as the major component of efflorescence.

Causes of rising damp

As we have seen, it is a natural process for moisture to rise from the ground in a porous medium such as a masonry wall. This has been recognised for many years, and the need for a damp-proof course has been a stipulation in public health legislation for well over a century. In building regulations, it is a requirement that any wall shall not transmit moisture from the ground to the inside of the building, or to any material in its construction liable to be adversely affected. The appropriate provision to satisfy this regulation states that the wall should have a damp-proof course 150 mm above the outside ground level. In addition, it is necessary to protect vulnerable timbers from dampness, and any sub-floor voids must be adequately ventilated.

Many buildings have had their dpc's rendered ineffective by bridging. This can take many forms, and Fig. 7.5 illustrates some of them. In particular, high external ground levels caused by paths or garden flower beds built up against the wall are common causes: hence the reason for the 150 mm height of the dpc from ground level. Abutting walls are likely to be troublesome unless they are constructed with a dpc and have another one at the top of the coping. Alternatively, abutting walls should be isolated from the main wall by a vertical dpc. External renders and plinths are often installed in an attempt to reduce dampness, but if they bridge the dpc they are liable, in time, to crack off the underlying brickwork or masonry sufficiently to allow moisture to rise in the crevice formed behind.

Internally, bridging is often caused by a floor screed being laid either without a membrane or including a membrane that is not overlapped with the dpc in the wall. Internal plaster may be applied to a wall in such a way as to bridge the dpc. In cavity walls, mortar droppings in the cavity can cause the dpc to be bridged. Cavity wall insulation can exaggerate the effects of such cavity bridging.

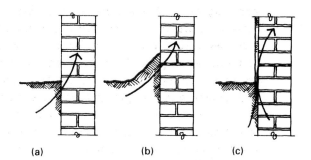

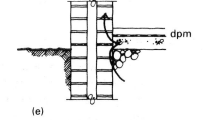

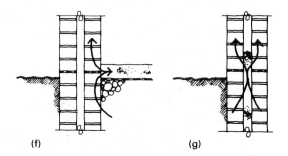

Fig. 7.5 Some causes of rising dampness in walls (after BRE Digest 245):
(a) no dpc, or existing dpc ineffective; (b) high ground level or path; (c) render;
(d) pointing unsuitable or dpc defective; (e) solid floor bridging dpc; (f) solid floor
bridging dpc; (g) debris in cavity.

Diagnostic procedures to help confirm the presence of rising damp are described in Chapter 9 and in Appendices F, G and H.

Remedial treatment of rising damp

As has been shown earlier, rising damp may develop if there is no effective dpc in a wall, and the other necessary conditions, such as a high soil moisture content, are present. One or a combination of therapies may need to be undertaken. First of all, it is possible that by providing field drains around the building, rectifying defective drainage, and promoting rapid drying-out of the affected walls, the problem will be resolved. Second, if the problem is related to bridging of the dpc, this must be remedied by conventional building methods. As a last resort a new dpc may need to be installed in situ.

Several techniques have been developed for the installation of a retrofit dpc in an existing wall. The use and effectiveness of each depends on the extent of dampness and on the form of construction involved.

Retrofit dpc's can be classified into three broad groups:

- ❏ *Physical damp-proof courses:*
  - brick replacement method;
  - saw-slot method;
  - Massari method;
  - atmospheric siphons.
- ❏ *Electrical damp-proof courses:*
  - electro-osmosis ('passive' or 'active' types).
- ❏ *Chemical damp-proof courses:*
  - pressure injection;
  - gravity transfusion;
  - injection mortar.

Physical damp-proof courses

Brick replacement method

Several types of retrofit physical dpc have been developed. An early method relied on the replacement of about three courses of bricks at the appropriate level (up to at least 150 mm above outside ground level). The new bricks must be at least Class B engineering brick quality, and should be laid in staggered bays (similar to underpinning) not exceeding 900 mm in length using a 1 : 4 cement : sand mortar. This method may be suitable where only a partial retrofit installation is required: for example, to the outer leaf of a cavity wall along one elevation. It is more difficult and expensive to apply this type of remedy to a long and thick wall.

Saw-slot method

In 1990 a De Beers Industrial Diamond Division Press Release (18/90) claimed that one of the most popular contemporary saw-slot methods was the HW process. The method is named after its inventor, Mr Huckendubler, and was described in that press release as follows:

'Ribbed sheets of 4016 A1S1 430 stainless steel are used, 1.5 mm in thickness and between 300 and 400 mm wide. The sheets are forced into the structure with a pneumatic hammer, capable of exerting a pressure of 20–40 kg, operating at a rate of 1000–1450 strokes a minute. The hammer is mounted on a framework fixed to the wall, which allows the tool to be moved to a horizontal plane.

On the inside wall, the metal sheets continue right through to the decorative coating, thus protecting the whole thickness of the wall. On the outside, they are left flush with the surface or slightly recessed, and the gap made good with waterproof cement.

An example of the HW process could be seen in work carried out in a suburb of Paris around 1990, where an old two-storey building required a retrofit dpc.

Sawing was carried out with a Longyear HE 630 machine mounted on rails fixed to the walls and driven by a 29 hp Volvo motor fitted with a safety cut-out. The 900 mm diameter blade was run at a speed of 47 m/s and cut to a depth of 500 mm. Coolant was supplied by a hydraulic pump, driven by a 32 A motor connected to a 38 V mains supply. Both electric and hydraulic controls are grouped together on a remote console. The job was completed in 2 days, with a total of 36 linear metres being cut. Two teams, each of two men, were employed, one to make the cut, the other to position the stainless steel sheets.

The advantages of diamond sawing are that it is quick, does not create vibration to weaken the structure, and the cut is made neatly and cleanly, thus facilitating the positioning of the steel sheets. The work can be carried out on floors, walls and ceilings, horizontally, vertically or at an angle.'

Other systems relied on cutting a slot in a convenient and appropriate mortar course, starting at a corner or at a door jamb. When the cut reached the inside it was cut right through the wall (Fig. 7.6). The slot was cut with one of a variety of tools: a hand saw, a chain saw or a power-operated circular or reciprocating saw of mild steel with tungsten carbide or stellite-tipped teeth. The BRE developed a modified chain saw suitable for use on mortar, using tungsten carbide cutting blocks brazed onto the links supported on a mild steel buttress (described in BRE Digest 27). The slot cut was about 7 mm wide.

Another method of cutting the slot uses grinding discs of glass fibre impregnated with carborudum. A protective shield round the disc has a vacuum extractor attached (BRE Digest 27).

In all these methods, a slot is cut about 500–1000 mm along the wall, and the new dpc is inserted. In thick walls (i.e. over 225 mm thick), usually each side of the wall is cut separately, but great care needs to be taken to ensure that the two slots are aligned with each other. Various

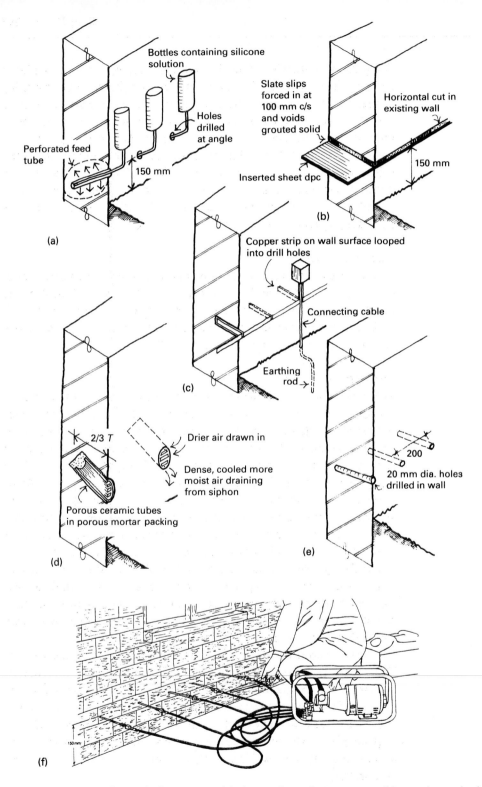

Fig. 7.6 Some typical retro-fit dpc systems: (a) chemical transfusion system; (b) saw-slot method; (c) electro-osmosis; (d) atmospheric syphons; (e) frozen siliconate pellets; (f) pressure injection of a chemical damp-proof course (Sovereign Chemical Industries Ltd).

damp-proof materials are used, including copper, bitumen or plastic sheeting. The slot must be filled with the membrane either by using several layers or by laying the damp-proof course material over a fresh bed of mortar. It is often difficult to obtain a satisfactory layer of mortar on both sides of these dpc's, and a system has been developed to overcome this difficulty. Particular care has to be taken to ensure that water, soil or gas pipes are not damaged and electrical wiring is not affected during the cutting work. Overall settlement of the building can be prevented during the application of the dpc by inserting it in alternate bays and using wedging slips of slate.

Massari method

A more recent system, used in conservation work, is the Massari method, named after Drs G. and I. Massari in Italy. This method involves the insertion of a polyester resin in a cut through the full thickness of the wall to form a damp-proof course that structurally integrates the wall above and below the cut (Ashurst & Ashurst, 1988).

Atmospheric syphons

The use of porous earthenware, clay or other ceramic tubes to assist in drying-out of wet walls suffering from rising damp is a very old-established method, developed in the nineteenth century (Ashurst & Ashurst, 1988). In this system small (approximately 50 mm diameter) tubes are inserted externally into holes made in the affected walls. They are generally installed at an upwards angle between 10 and 15° to penetrate about two-thirds to three-quarters of the thickness of the wall, at about 300 mm nominal centres (Fig. 7.6(d)). They are embedded in a salt-free sand, which acts as a porous packing mixture, and the holes are faced up with a suitable mortar incorporating a ventilating grill.

The theory behind this system is that the tubes are supposed to attract water to their evaporating surfaces by capillarity, drawing dampness from the inner damp surrounding masonry. The humid air in the tube would fall out of the tube and be replaced by air, and the drying process would therefore continue.

There are, however, a couple of inherent problems with this method. There is, for example, a strong likelihood that the pores on the surface of the tubes will become blocked as mineral salts accumulate during the evaporation process. The accumulated salts would block the drying pores and also contribute to the general dampness as a result of their hygroscopicity. Moreover, because the tubes are capable of attracting water capillarity, they will inevitably be reluctant to lose that water by evaporation.

No laboratory tests have been reported that demonstrate the effectiveness of syphons conclusively. Furthermore, because no satisfactory theory supports their mode of operation, they have not received any official approval as a method of dealing with a serious rising damp problem.

Electro-osmosis

Electro-osmosis was developed as a retrofit treatment of rising damp electrically in 1930 by the two Ernst brothers in Switzerland (Richardson, 1991). Their first type was an active system, in which a d.c. supply from a mains transformer passes into wires that connect up with electrodes in the wall. One of the main problems with this method was that the metal anode was depleted by electrolytic corrosion. However, platinised titanium and other resistant materials metals have largely overcome this problem (Ashurst & Ashurst, 1988).

The second electro-osmosis method, and one more widely used in the UK, is the passive system, in which electrodes are inserted into holes drilled in the wall at an appropriate height, and these are then wired up to the earth. The electrodes installed in damp walls have an electrical potential relative to earth, which can induce rising damp. Richardson (1991) reported that the dampness could be prevented by connecting or shorting-out the electrodes, or connecting wall electrodes to earth, in order to remove this potential.

In either case the electrodes are intended to provide the electrical charge that repels the charged water molecules as they rise in the wall from the ground (Fig. 7.6(c)). The short-circuit should thus induce drying of the wall.

Electro-osmosis is, in fact, a well-established process for drying clay soils, where positively charged water molecules become attached to the surfaces of the minute clay particles, which are up to 0.002 mm (i.e. 2 μm) in size. By applying electrical voltage to the clay, the positive ions in the water are attracted to the cathodes of the electro-osmotic system, and the clay particles move towards the anode.

However, electrical charges in rising damp vary widely depending on the types of salt present. These salts, as we have seen, may originate in the soil, or they may be a constituent of the masonry units or mortar, and their presence and extent may vary according to their position along or up the wall. To be effective, therefore, an electro-osmotic current will have to be varied as appropriate along the wall, and this would appear to be impracticable. It has also been suggested that, as rising damp may vary with the season and the height of the water table, the electro-osmotic electrical system might then be 'broken' during drier periods, and then might not be re-established again during the next damp period. Another problem with this method is that the potential is sometimes reversed relative to the direction of the water flow (Richardson, 1991).

In addition, it has proved difficult to demonstrate convincingly the effectiveness of passive systems in the laboratory. As a result, the use of this version has declined in recent years. Moreover, there is still a lack of evidence that any electro-osmotic system has been significantly successful. For these reasons, electro-osmotic systems have not been given official approval by either the BRE or the British Board of Agrément (BBA).

Chemical damp-proof courses

General

The use of a chemical damp-proof course to control and prevent rising dampness has become by far the most popular system in the UK. It was first used in the UK in the early 1950s (Howell, 1994). There are several different systems of application, and the two main groups of chemicals are in either solvent-based solutions or water-borne products. Examples of all the main variations have been tested independently and have been approved by the BBA. Their application is included in BS 6576 (1985).

The main types of chemical employed as the active ingredients in chemical damp-proof course injection fluids are as follows:

❑ Organic solvent-borne chemicals:
  ■ silicone solutions;
  ■ aluminium stearates.
❑ Water-borne (aqueous) chemicals:
  ■ siliconates (water-soluble silicone compositions);
  ■ micro emulsions.

Application is usually either by pressure injection or by a gravity transfusion system.

Mode of action

It will be recalled that water wets and spreads along the surface of the pores naturally present in the masonry. The surface tension forces at this interface can be modified by using water-repellent chemicals.

One of the most familiar use of water-repellents (Fig. 7.1(b)) is in the shower-proofing of wet-weather clothing. The visible effect here is to prevent rainwater from penetrating into the clothing by repelling it. Water droplets can be seen on the surface of the garment, and many of them will run off the clothing rather than penetrate into the weave of the material.

If a capillary surface is coated with a water repellent, the contact angle of the water with the surface (Figs 7.1(a), 7.1(b)) will be changed from a small angle (spreading) to a large one (repellence). This effect can be produced on many masonry materials, including brickwork, stone and mortar, all of which are naturally hydrophilic and are easily spread over by water. By creating this high contact angle, the meniscus in the capillaries reverses and the capillary forces operate in the opposite direction: that is, downwards in the case of rising dampness.

The injection fluids are believed to function by depositing their active ingredients (e.g. silicone or stearate from aqueous or organic solvent) onto the walls of the capillaries in the masonry, where they act as pore liners. In the case of solvent-borne solutions, they displace water present in the wall, and when the solvent evaporates the resin is left to form a permanent barrier to rising damp. These active ingredients may then polymerise and cross-link with each other and with surface water molecules and hydroxyl groups, to form permanent water-resisting layers. It is possible that a full chemical bond to the masonry will not be so well developed with stearates, and this might lead to subsequent elution if contamination of the wall with detergent occurs.

In general, the active ingredients do not block or fill the pores. Rather, they coat the capillaries of the masonry, and their main effects are those of water-repellence. Under certain circumstances, however, pore blocking can occur with some injection fluids.

Silicones

These are almost always pressure-applied as organic solvent solutions. The active ingredient, commonly 50% silicone in white spirit solution, is in the form of partially hydrolysable ethoxy groups, which react with water in masonry to form silanol and siloxane groups and, ultimately, fully condensed durable resins.

Stearates

Stearates, or more correctly polyoxoaluminium stearates, consist of aluminium oxygen chains with organic stearate residues attached, and these latter, being hydrophobic, provide the water repellency. They are almost always applied as organic solvent solutions and, like silicones, polymerise on contact with water. The stearate solution has a lower viscosity and so appears to penetrate particularly well when pressure-injected into brickwork.

Siliconates

These are always applied as aqueous solutions of sodium or potassium methyl siliconate. The solution reacts with atmospheric carbon dioxide after application to the masonry, and forms a water-insoluble siliconate resin of high molecular weight, which provides water repellence. The sodium carbonate by-product from this reaction sometimes causes efflorescence. Therefore some formulations use potassium siliconate, because potassium carbonate does not cause the same problem. Siliconates are normally formulated as a solution in water with a 50% active ingredient.

A special formulation of siliconate has been developed in which a catalyst is mixed with the solution at the time of application to speed up the curing process, and to ensure that the injected chemical stays close to where it is applied in the wall, and does not migrate or get leached from the application site by active rising damp, for instance.

Micro emulsions

A more recent development in chemicals for damp-proofing is the use of micro emulsions. These water-borne solutions are usually concentrates based on silanes and oligomeric alkylalkoxy siloxanes. They are highly viscous, and have smaller particles than other dpc chemicals. As a result, micro emulsions, the manufacturers claim, can achieve a more even distribution, more rapid spread and better penetration of the active silicone throughout the wall.

In addition, micro emulsions are more environmentally friendly than organic-solvent-based dpc fluids. Unlike the latter they leave no lingering smells on application, and are non-flammable. Other problems associated with solvents when used in a confined space, such as dizziness and nausea from the noxious fumes, are also avoided. BBA certificate 95/3123, for example, is one of many that cover such products.

Comparisons

There are few published data to enable an up-to-date comparison to be made between the various active ingredients in damp-proofing injection fluids. Most have been in use for many years, and have been approved by the BBA following both laboratory and in-situ field trials. However, variations in both the consistency of the masonry being treated and the chemical being used will, along with the quality of workmanship, markedly affect the efficacy of the treatment.

As has been acknowledged by the chemical damp-course industry, there is no one simple test for detecting the type of chemical used that is fully inclusive and of universal application (see BWPDA publications in Bibliography). Clearly, these are areas that would benefit from much more independent research.

It has been suggested that the silicone repellents in common use for damp-proofing will not perform well in masonry containing highly alkaline conditions. For example, such conditions might occur in limestone masonry, or in newer constructions containing relatively fresh cement mortars and renders, where the pH value can be as high as 11. It is believed, however, that the different forms of silicone resin produced by different manufacturers will vary in their performance under highly alkaline conditions, but there is little independent information on this aspect. Still, it can be stated with confidence that aqueous solutions will not be subject to durability problems in such masonry, because they are exceedingly alkaline in the first place.

Solvent effects

There are basically two types of solvent: water and organic. Technically, emulsions are suspended in water, which acts as a carrier, but the products are water based. Thus these dpc fluids are either solvent based or water borne.

The efficacy of the chemical injection system will depend on the extent of penetration of the injection fluid through the damp wall. This is clearly going to be influenced by the character of the injection fluid solvent. If the solvent is miscible with the dampness in the wall, as is the case with a water-based siliconate, a diffusion effect might be expected. If it is immiscible, as with an organic solvent-based fluid, such diffusion would not occur.

When an organic-solvent-based injection fluid is applied to the wall by a pressure system, it has been suggested that the invading chemical will not remove the residual dampness by driving it all before it. Because of the variability in porosity across a wall and in the constituent parts of the masonry, some parts of the fluid will advance more than others: a process that has been described as 'viscous fingering'. This occurs in areas of greater permeability, and ultimately pockets of residual dampness will become surrounded by the injection fluid. These pockets remain in situ however long the injection process continues, and might constitute a path for continued rising damp, especially in very wet and heterogeneous construction. It has been proposed that, to reduce this fingering problem, lower pressures and more viscous fluid should be employed.

To some extent the uneven nature of penetration of fluid applied under pressure will still occur with water-based systems. But these systems have the advantage that diffusion of the active ingredient into the damper areas can take place subsequently, which might in the end provide more complete penetration of the wall. These diffusion processes are slow, and it has been suggested that the water repellent might be carried up the wall some distance before the curing process with the carbon dioxide can be complete. Clearly, the extent of post-treatment diffusion will depend on the speed of curing.

These theoretical objections to the scientific basis of the effective functioning of chemical damp-proof courses have been gaining increased credence (Rowland, 1996). Nevertheless, they must be viewed in the context of the large number of apparently successful installations that are in use. Moreover, it may be that the excessive heterogeneity and very high moisture levels do not occur very frequently in practical damp-proofing.

Other more practical effects of organic solvents concern fire risks with flammable solvents, their greater smell and increased cost. They can also act as a solvent to bituminous-based water barriers and polystyrene insulation in the vicinity of newly injected chemical damp-proof courses. A summary of the comparative effects of organic solvents and aqueous solutions and their properties and uses can be seen in Table 7.4.

Scientific evaluation of chemical damp-proof courses

Research background
In comparison to other areas of study in the built environment (e.g. architectural acoustics, thermal transfer in buildings, drainage fluid flow), the extent of scientific work on dampness has not been as intensive. For example, there has been relatively little independent research into the efficacy of either retrofit chemical dpc's or hand-held electrical conductance-type meters (Howell, 1995). Some of the work that has been undertaken has indicated that the accuracy of diagnoses and the efficacy of remedial treatments may be doubtful (Houston, 1990; Howell, 1995).

Overall effectiveness of chemical dpc's
Generally, however, the public perception of chemical damp-proof course systems is that they are effective. This has been gained as a result of a combination of good marketing by the chemical damp-proofing industry and many prima facie successful commercial treatments in numerous buildings over the years. There is a theory that explains how

Table 7.4   The effects of solvents on the properties of damp-proofing injection fluids.

Property	Organic solvents	Aqueous (water based)
Cost Fire risk } Smell	Higher	Lower
Leaching from application site	None	Possible, but can be fixed with a catalyst, which speeds up the polymerisation of the silicone after injection
Safety risks	Fire Fumes/smell	Strongly alkaline solution, which can cause damage to the skin and eyes
Treatment	Visibly penetrates better into masonry (especially stearates)	More viscous, does not visibly penetrate so well, but may diffuse in wet walls
Effects on bituminous materials	Acts as a solvent, causing them to dissolve	No adverse effect
Effects on polystyrene	Acts as a solvent, causing them to dissolve	No adverse effect

they work, but there is still a shortage of independent test data showing how effective they are either in the laboratory or in the field. The long-term efficacy of retrofit chemical dpc's has still to be ascertained. Thus much more work by impartial research agencies is required to ascertain the durability of these systems.

Some studies (Houston, 1990), for example, have suggested that for certain walling, such as composite sandstone wall construction found in Scotland, chemical injection may not be as effective as was first antici-pated. Key factors in the efficiency of the penetration of the chemical, and thus its success as a damp-proofing solution, are:

❏ the duration of the injection application;
❏ the spacing and depth of the injection holes;
❏ the suitability of the pressure employed;
❏ the type and quality of chemical used;
❏ the extent of large voids within the core infill;
❏ the quality of construction itself.

Moreover, support measures such as lowering the outside ground level, an appropriate replastering system, repairing defective walling and rainwater goods, will be required to ensure a successful outcome. Indeed, these latter measures in themselves may be sufficient to control an apparent rising dampness problem.

In fairness, however, the British Wood Preserving and Chemical Damp-Proofing Association (BWPDA) has commissioned some work in this area. It has published a series of short but informative technical information papers on a variety of topics related to dampness assessment and chemical damp-proof courses (see References).

Laboratory assessment methods

In the laboratory, many attempts have been made to construct walls or columns that, under controlled conditions, develop rising dampness. It is desirable to test both the masonry units and the mortar together. In addition, the high alkalinity of new cement and lime-based mortars may affect the performance of some of the injection fluid, which could distort the results.

A close approximation to the constitution of well-weathered lime mortar was required, and research at the BRE was directed initially towards developing and testing a suitable mortar. A suitable composition was found that consisted of chalk, lime, ground brick pozzolana, and graded and washed sand. In subsequent trials, rising damp was established and the effectiveness of an injected 5% silicone solution was demonstrated in tests carried out by Sharpe at the BRE in the late 1970s (Table 7.5).

Table 7.5 Laboratory results of a test of a chemical dpc.

Unit no.	Initial moisture content (%)	Moisture content (%) 11 weeks after injection treatment
Brick 6	0	0
Brick 5	5.0	0
Brick 4	8.0	0
Brick 3	10.5	0.9
Brick 2	14.9	3.8
Brick 1	17.7	16.5

Note: The base of the brick column is placed in a tray with 20 mm of water.

The results in Table 7.5 are only of one range of samples from tests carried out in the late 1970s. As these results were from a small sample of tests taken many years ago, they should not be taken as typical. What they do suggest is that, provided the walling is in good condition and the injection is undertaken in controlled conditions, the chemical retrofit dpc can prove effective. A much more wide-ranging series of laboratory and field tests is needed to investigate the efficiency of such treatments for the variety of wall constructions and subsoil conditions encountered.

Joint laboratory and field trials

By the early 1980s, the BBA had introduced an assessment scheme for chemical damp-proof course injection fluids. This was based on laboratory assessment using methods similar to the BRE approach, together with field examinations in which drilling samples were analysed for moisture content to illustrate the control of capillary rising damp. These drilling samples were also analysed for salt content in an attempt to prove the previous presence of rising damp in the wall under investigation. Most of the commercial damp-proof course injection fluids based on siliconates, silicones and stearates have been approved by the BBA. Independent approval of this type is a requirement for chemical damp-proof courses applied under BS 6576.

In Belgium, at the Centre Scientifique et Technique de la Construction (CSTC, 1985), research was carried out on the effectiveness of the main retrofit dpc's used in Europe. It was generally found that physical dpc's performed best, followed by the various chemical dpc's, with electro-osmosis and atmospheric syphons being the least effective.

Application

The installation of chemical damp-proof courses is covered by BS 6576: 1985 (and BWPD Code of Practice, 1986).

Siliconates, or to give them their full name, sodium or potassium methyl siliconates, are the active ingredients in aqueous injection fluids (i.e. they are water based). They are applied by gravity feed (also known as transfusion), by pressure injection, or as frozen sticks, although this latter method is not included in BS 6576.

Gravity feed or transfusion system

This system aims to apply a measured quantity of fluid to the wall through a series of holes normally up to 20 mm in diameter, which are drilled at spacings not exceeding 175 mm. The holes are drilled to within 40 mm of the far structural face of the wall in a convenient mortar course (Table 7.6). The holes are usually drilled horizontally, but in brickwork 115 mm thick (for example, with internal walls or the leaves of a cavity wall), they can be drilled at a downward-sloping angle of 45° and at

115 mm spacing, and should form a line for the new damp-proof course 150 mm above external ground level.

The diffusion fluid is applied by filling a container that feeds by gravity the perforated metal tube, fitted with sponge rubber washer, inserted into the wall. It is essential to ensure that the fluid diffuses into the wall, and is not lost through cracks or fissures in the wall. This is usually prevented by the use of the sponge rubber washers around the inserted part of each tube. The quantity of fluid in the bottle is calculated to be sufficient to treat adequately a particular type and thickness of masonry.

Frozen sticks diffusion system

Another method of application of siliconates involves the insertion of cylindrical pellets of frozen siliconate solution into 22 mm diameter holes, which are drilled in the mortar bed at 110 mm intervals at 150 mm above outside ground level. As the pellets melt, the active ingredient siliconate diffuses into the surrounding masonry, saturating it. The performance is similar to that of the transfusion system.

In an Agrément Board certificate (92/2849) that covers this process, it is stated that siliconate pellets are not suitable for use in new construction, where the mortar will be highly alkaline.

Pressure injection systems

These can be divided into high-pressure and low-pressure systems, and their use is dependent on the types of chemical used, as shown in Table 7.6. Pressure injection of aqueous solutions of siliconates is usually by the low pressure system. The fluid is pumped into predrilled holes, normally in a convenient mortar course in the wall, usually 9–16 mm in diameter at 150–175 mm centres. The holes are either horizontal or at a downward-sloping angle, terminating in a mortar bed joint at the level at which the dpc is required, at least 150 mm above outside ground level (Fig. 7.6(b)). Solid walls up to 450 mm (18 in) in thickness can be treated from one side according to BS 6576.

The dosage rate is about 1.5–2 litres per metre run of 225 mm thick wall, and pro rata for thicker walls (Kyte, 1987). Enough time must be allowed for all the fluid to flow from the container and to be absorbed by the masonry, and the process is usually continued until the fluid is seen to penetrate to the surface of the brick or mortar. This may take from a few hours to a day or more.

Injection is carried out using a lance with an expanding seal that closes the aperture of the holes drilled in the wall during the injection process. It is important to ensure that the pressure is maintained during injection. A rapid reduction in pressure will indicate a loss of fluid through a crack or hole in the masonry. The precise volume of fluid to be injected will depend on the wall thickness and the type of masonry, and the period of injection into each hole is usually timed.

Organic-solvent-based injection fluids usually have either silicone or stearate as the active ingredient. These fluids are nearly always applied by high-pressure injection (Fig. 7.6(f)) using an injection nozzle. One or more nozzles may be used at a time, but special care must be taken to avoid blow-outs or sudden losses of pressure as a result of contact with a fissure in the masonry. Ideally, injection should be continued until the fluid is seen to exude back out of the masonry to form a continuous band along the line of the injection. However, when the masonry is not sufficiently porous, or if it cannot be seen, then a time period for injection is adopted.

Normally, for walls greater than 120 mm, injection is carried out from both sides of the wall. This avoids losing fluids in the vertical joint behind the facing brick or stone.

Injection holes in high-pressure systems are between 9 and 16 mm in diameter, and are drilled at sufficiently close intervals to ensure that the injected dpc will be continuous. The holes are drilled usually in the masonry units if they are sufficiently porous, and a pattern of two holes per stretcher or one per header is usual. Mortar line injection is commonly adopted when the wall is constructed of very impervious bricks or stones. It is often difficult in these circumstances to obtain very high pressure, because the weak mortar commonly found will not normally withstand the pressurised injection fluid. In such cases, repointing may have to precede injection.

When walls thicker than 120 mm are encountered, the double drilling technique has to be employed. In this process the nearest face is first injected, and the wall is then redrilled to a sufficient depth to treat the inner section of wall, generally proceeding at up to 150 mm increments in depth. If horizontal drilling is to be used, the existing holes can be redrilled to the greater depth required. If drilling is in a downward direction finishing near a mortar bed joint, a separate hole might be drilled for the interior injection.

Once the injection fluid has achieved optimum penetration, the exposed drill holes should be plugged with a 1 : 5 cement : sand mortar.

A summary comparing the characteristics of high-pressure and low-pressure systems is presented in Table 7.6.

Safety precautions in chemical damp-proof installations

Some damp-proofing chemicals are based on organic solvents of the white spirit type. The vapour that is produced when such installations

Table 7.6 Summary of characteristics of pressure injection systems.

Characteristics	Type of system	
	High pressure	Low pressure
Pressure range	0.7–0.9 MPa (100–130 lb/in^2)	0.15–0.3 MPa (20–50 lb/in^2)
Chemicals used	Organic-solvent-based silicones or stearates	Aqueous solutions of siliconates or micro emulsions
Wall thickness	Walls up to 120 mm thick: drill from one side only Walls greater than 120 mm thick: drill from both faces of the wall, or by progressive injection	Walls up to 450 mm thick: drill from one side only to at least half of the wall thickness Walls thicker than 450 mm: drill from two faces, or by progressive drilling from one side only
Suitability	Bricks, porous homogeneous masonry, and mortar joints	Mortar courses and more heterogeneous masonry

are drying out after application is classified as flammable, and smoking and the use of heating or cooking appliances with naked flames should cease, and not be recommenced for 48 hours after treatment. It is for these reasons that the water-based solutions are becoming more popular.

In any event, safe electrical equipment using 110 V should be employed.

In addition, suitable protective clothing should be worn during drilling and injection operations with all types of retrofit dpc fluids; protection of the face and eyes is particularly important. A risk assessment procedure may also have to be carried out prior to such work.

Storage and transportation of containers should be carried out safely and securely; containers should be correctly labelled and, when empty, disposed of in a safe way if they contain solvent residues.

Drilling patterns

When the use of chemical damp-proof course injection systems commenced in the late 1950s, application was either by hand into predrilled holes in the brickwork or by the gravity-feed system. In either case, application was to the mortar. Holes would be drilled at appropriate centres into a convenient and appropriate mortar course (Fig. 7.7(a)). When pressure injection was introduced in the 1970s, it was appreciated that much of the mortar would be in such a poor condition that it would not be possible to pressure-inject it. The alternative method of application proposed was to inject the bricks or masonry units themselves under pressure; the fluid would then penetrate into adjacent mortar via cracks and capillaries in the masonry units (Figs 7.7(b) and (c)).

The typical drilling pattern for the masonry porous units was a line of holes approximately 100 mm apart (i.e. two holes per stretcher and one per header (Fig. 7.7(c)). With less porous units, mortar line drilling was the only possible method. In some systems, a single line of injection holes was regarded as sufficient (Fig. 7.7(b)), but with others a double line with the intermediate vertical joint also being injected was the application method adopted (Fig. 7.7(e)).

All these techniques can be used on solid walls. As indicated earlier, however, their effectiveness on old solid walls with a middle core of rubble fill may not be as good as with more homogeneous masonry. This is because many of the infill layers may contain excessive voids or poor-quality material. On cavity walls both leaves should be injected. Where access is restricted, double drilling techniques can be employed.

Damp-proof courses on external walls should be at least 150 mm above outside ground level. If this requirement means that the new dpc is installed at or above the level of a suspended timber floor, then the structure must be isolated from any dampness, and should receive timber preservation treatment. Timber joists built into solid walls should be protected against damp by being covered with a joist shoe made from an impervious material (e.g. plastic). Injected dpc's should also protect suspended timber floors adjacent to internal walls from rising damp. In buildings with solid floors, the damp-proof course on internal walls should be as close to the floor as possible, and an arrangement should be made to tie in the wall dpc with any membrane in the floor.

The effectiveness of a horizontal damp-proof course can be reduced if dampness can rise in adjacent structures and penetrate into the 'protected' masonry. In such situations, a vertical damp-proof course should be installed (Fig. 7.8). Abutting structures such as chimney breasts, walls of adjacent properties, garden walls, and where different floor levels require damp-proof courses to be installed at different levels are all situations where vertical dpc's are necessary. A vertical damp-proof course isolating adjacent property should be at least 1200 mm high, and should extend at least 300 mm above the height at which the rising damp is recorded.

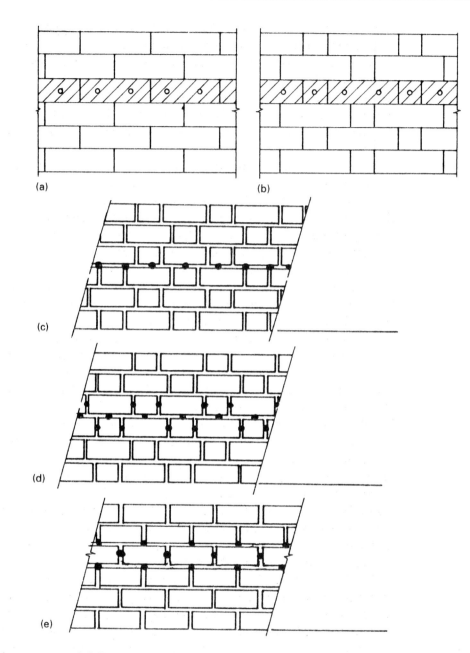

Fig. 7.7 Typical drilling patterns: (a) 115 mm wall or leaf; (b) 230 mm solid wall; (c) typical mortar line injection; (d), (e) alternative patterns for mortar line injection.

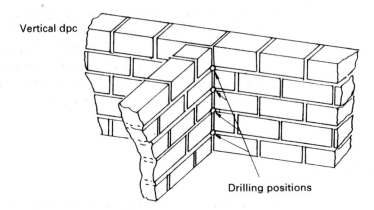

Fig. 7.8 Vertical damp-proof course (to prevent internal moisture transmission) achieved by chemical injection.

Injection mortar

Injection mortar is a type of retrofit chemical damp-proof course in which a recently made-up cementitious slurry is inserted into holes drilled in the masonry, either in the units or in the mortar.

Agrément Certificates 93/2870 and 95/3123 cover two typical injection mortar systems. The process involves the introduction of a slurry of the proprietary injection mortar into 19 mm diameter holes at 115 mm centres and at a downward angle of 20–30°. The depth of the holes should be equivalent to the thickness of the walls.

Most injection mortars are believed to be Portland cement based, with a proportion of fine silica sand and chemical additives. It is suggested that the damp-proofing effect is achieved by the formation of insoluble calcium silicate salts as a result of the reaction of quartz with free lime in the masonry. These salts are believed to block the pores in the vicinity of the injections.

It is claimed that injection mortar is particularly successful on rubble-filled or flint walls, and it is relatively easy to apply in such constructions.

Durability of chemical dpc's

As has already been mentioned, numerous retrofit chemical dpc's have been used for over 40 years in many parts of Europe. Some indication of their durability therefore is now emerging. Most suppliers of chemical dpc's anticipate that their products should remain effective for at least 20 years. In some cases a 30-year guarantee may be available; but as with all such guarantees, they can be difficult to enforce in the long run.

The long-term performance or actual life expectancy of these materials has, however, not been established within measurable limits. It is not clear whether existing installations may need to be retreated at some future period. Still, there have been no reports of significant failures with these types of retrofit dpc.

Replastering

The major visible effects of rising dampness are, of course, caused by the deposition of salts on wall surfaces where evaporation of water is taking place. These hygroscopic salts attract moisture from the atmosphere, and towards the apex of the rising damp this hygroscopic moisture is the major factor in the total dampness situation. It will continue to be a problem even after the rising dampness has been eliminated.

Therefore, for any rising-damp curative treatment to be satisfactory, the problem of salt-contaminated plaster must be addressed, and this normally means removal. Ideally this would be delayed for as long as possible after the installation of the retrofit dpc so that any residual salts that accumulated at drying surfaces during the first drying-out period could be eliminated at the same time. It is usually estimated that masonry will dry out at a rate of 25 mm per month, so a 225 mm (9 in) solid brick wall would be expected to dry out by about nine months after installation of the new DPC. A nine-month delay in replastering would therefore be preferred in that situation. In practice, of course, it is only very infrequently that such delay can be encountered. Rapid drying-out methods listed in BRE Digest 163 could be used to accelerate this process.

The purpose of the plaster is twofold: (a) to replace the existing contaminated plaster, and (b) to provide a barrier to prevent any residual dampness and hygroscopic salts from reappearing on the newly plastered surfaces. The new plaster must act as an effective barrier, and this is achieved by correct formulation as well as by proper application. The extent of replastering must be sufficient to prevent any residual moisture and contaminating salts in the wall from bypassing the new barrier. Normally a distance of 300 mm above the last sign of rising damp is specified. At dpc level itself, the new plaster should not cause bridging. Usually this could be achieved by stopping the plaster short of the floor and arranging a gap of about 25 mm, which can, in turn, be concealed by a skirting board.

A vital requirement of the new plaster is that it must not contain gypsum (calcium sulphate), because this will quickly deteriorate if it gets wet, either directly or by contact with hygroscopic salt moisture. Therefore plasters used for replastering are usually cement based (but see precautions below). One of the most successful formulations (Table 7.7) is to use a 3 : 1 sand : cement mix, provided coarse-graded (salt-free) sand (Table 3.6) is employed; sands graded to BS 1199: 1986 Type A or Type M in BS 882: 1983 are suitable. This has proved to be a most effective barrier even where positive hydrostatic pressure is present, and the protection can be enhanced by incorporating waterproofing admixtures (or additives) such as stearates, oleates or styrene butadiene. These admixtures act as air entrainers as well as improving workability. They reduce the amount of water needed for mixing, and consequently reduce drying and shrinkage cracking.

The disadvantage of this type of plaster is that it is dense and strong, and may therefore attract condensation. This can be partially controlled by designing and applying a top coat of a more porous composition. As it is a strong mix, there is always the risk of breakdown or detachment on

Table 7.7 Comparison of replastering systems.

Factors	Types of plaster	
	Sand : cement	Renovating plaster
Thermal resistance 15 mm thickness (m^2 °C/W)	0.021	0.083
Application First coat Second coat Third coat	3 coat work (23 mm)[a] 3:1 and WR[a] 10 mm 4:1 10 mm Sirapite 3 mm	2 coats (12.5 mm) Up to 11 mm Up to 1.5 mm
Resistance to Hydrostatic below ground pressure Vapour permeability Condenstion and mould growth	Can be used with modification Less More likely	Not suitable More Less likely
Colour and texture	Dark greyish with rough finish	Pinkish or light greyish with smooth finish like gypsum plaster
Suitability	Brick and cement mortar substrates	Stone and lime mortar substrates

[a] Integral waterproofer commonly added to improve water resistance.

weaker masonry backgrounds. This does not seem to be a major problem in practice, however, possibly because thermal and moisture movements are low or near ground level in a wall, as their conditions are more stable.

Nonetheless, with old buildings of solid masonry construction, care should be taken in selecting an appropriate rendering system. It has been pointed out by Howell (1995) that replastering such walls with a strong, impermeable sand-and-cement render is inappropriate, for a number of reasons. First, most old properties were built using lime mortar, which as we have seen is softer and more porous than its cement counterpart. Cement coatings or mortars are not compatible with substrates containing lime in the mix. Second, solid masonry buildings, being basically porous, rely on the 'overcoat' principle and, by virtue of evaporation on both sides (Table 2.1), will be in dynamic equilibrium with the moisture ingress.

Accordingly, since the late 1970s, the use of renovating plasters as an alternative to sand/cement mixtures has been widely adopted. These are premixed cement-based lightweight plasters. They are thus based on a mixture of cement, lime and an inert aggregate such as perlite, and incorporate an integral waterproofer with salt retardant and plasticiser additives.

Renovating plasters are easier to apply (normally in two coats), and are quicker drying, have better thermal insulation properties and, being weaker mixes, can tolerate more movement in the background. The plaster is more porous to water vapour, and so drying out of the wall is facilitated. It is the preferred replastering system where there is a likelihood of condensation. Condensation is lower because of better thermal insulation properties, and as the surface has a high alkalinity (pH up to 12) mould growth is inhibited. These renovating plasters would appear to have considerable advantages over traditional sand : cement renders in situations where rising dampness has occurred, but in practice there have been reports that they do not always perform well under severe conditions. Several renovating plasters have BBA certificates (e.g. 95/3123 and 89/2299, to name but two).

Dry-lining

In some cases it may not be feasible to replaster walls after a retrofit dpc has been installed. For example, the wall surfaces may be extremely friable or the wall profiles may be uneven.

Normal dry-lining methods are unsuitable on walls that are persistently damp because of the presence of high concentrations of hygroscopic salts. Indeed, British Gypsum (1991) advises against the use of ordinary Gyproc plasterboard for use in areas subject to continuous damp or humid conditions. Special dry-lining boards must therefore be used in such situations. This can be done in conjunction with a BBA-approved ventilated dry-lining system based on a high-density polyethylene (HDPE) membrane, which provides a vapour-resisting surface suitable for conventional plastering and/or dry-lining techniques. Boards with expanded polystyrene and especially polyurethane backing (25 mm minimum thickness) can act as an effective vapour check as well, provided the joints are sealed in the correct fashion. Such boards should be directly fitted mechanically, or sometimes on to pre-treated battens. It is not appropriate to use gypsum adhesives in 'dot and dab' applications directly onto the wall surfaces.

As an alternative to replastering walls where rising damp has occurred, dry-lining has many advantages. In Scotland, for example, it is a common system for covering the internal surfaces of external walls. Plasterboard, or in older buildings lath and plaster, is fixed to timber battens, which in turn are usually fixed to the wall. A dry-lined wall provides a good surface for finishing and, by lowering the heat capacity of the internal surface of external walls, would reduce condensation risks. It has the additional effect of concealing rising dampness in external walls until such time as severe deterioration and possibly decay of the timber battens has developed.

When dry lining is selected as the most appropriate construction for use on a wall where rising damp has occurred, the battens must be of pre-treated timber, and must be protected from any dampness in the wall as well by use of a damp-proof membrane such as polythene sheeting. Alternatively, vertical timber studs may be fixed to horizontal timber battens, which are fixed directly to the wall using protected metal or plastic fixings. Physical separation of the battens from the wall is essential if treated timber is not used, and the formed space must be ventilated.

The residual moisture and salts present in walls that have been affected by dampness may be highly corrosive to ordinary wire nails. Galvanised mild steel or, better still, copper nails should therefore be used. Fixings may puncture physical damp-proof membranes, but novel methods and design of fixings have been developed to minimise this problem.

The use of foil-backed plasterboard fixed directly onto walls that have been affected by rising damp has been proposed and used successfully over many years. It is fixed directly onto the substrate using stainless steel fixings and the boards are cut back to allow a 50–75 mm foil overlap at the joints. At the base of the plasterboard, the excess foil is wrapped around the lower surface of the board and concealed behind the skirtings. It is important to ensure that the aluminium foil or polythene backing is thick enough to withstand the corrosive salts in the wall. After fixing, the tops of the nails are covered with a spot of PVA emulsion, and the plasterboard joints are scrimmed prior to the application of a top plaster coat.

External rendering

As a remedial measure to prevent rising dampness, it is often necessary to ensure that external rendering does not bridge the damp-proof course. Various techniques can be used: the render can be undercut, stopped or finished in a bellcast mould (Fig. 7.9).

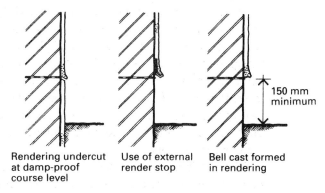

150 mm minimum

Rendering undercut at damp-proof course level

Use of external render stop

Bell cast formed in rendering

Fig. 7.9   Prevention of bridging of a dpc by external rendering (BS 5628 Part 3).

Drying aids

In situations where the walls have been treated with a retrofit dpc and are still likely to be subjected to persistent wetting, some additional precautions may be worthwhile: for example, the use of temporary and permanent drying aids. Dehumidifiers and warm-air heaters can be used as temporary drying aids. Permanent drying aids such as field drains (see also next chapter) can be used to enhance the evaporation from the base of the wall (Ashurst & Ashurst, 1988).

Cosmetic treatments

In certain situations it may not be possible or feasible to cure a rising dampness problem. For example, in some basement walls or other inaccessible parts of a substructure it may be impossible to either remove or repel the source of moisture. In such cases there may be a temptation to hide the effects of damp by using some form of masking or lining technique, such as plasterboard on battens fixed to the wall, or plasterboard on framing away from the wall. However, this approach has the basic drawback that any such remedies will reduce the rate of evaporation from a wall and allow the damp to rise further. The dynamic equilibrium between the moisture reservoir and moisture sink causes rising damp. This equilibrium is influenced by the rate at which water is fed to the wall via the soil and the rate at which it evaporates away via air flow across the surfaces. If it is disturbed, this could increase the height and intensity of the dampness in the wall. This could pose serious long-term dangers to timber components attached to or embedded in the affected wall.

Conclusion

Rising damp is usually not as extensive or as troublesome as other forms of dampness (Coote, 1975). In reality it is relatively uncommon, and can be easily misdiagnosed by surveyors (see Chapter 9). After all, it only has an effect on walls to about 1000 mm above ground level. Thus surveyors should take readings all the way up the height of the wall. Condensation, rainwater penetration and other sources of dampness can affect other elements of a building as well, and usually to a much greater extent. Granted, a rising damp problem could cause an outbreak of *Serpula lacrymans*, which could spread to other levels. But as has been shown earlier, rising damp is not the most prevalent source of moisture to trigger this most serious form of fungal attack.

For these reasons surveyors should be wary of recommending retrofit dpc installations cavalierly where rising damp is suspected. Further investigations may be needed to confirm that it is the actual rather than apparent source of the problem. Even so, it is possible that by reducing the water table of the subsoil around the building and allowing the affected walls to dry out properly, the problem will be curbed if not cured.

Chapter 8
Penetrating Damp

Background

Problems of penetrating damp

Because of the relatively high profile that condensation and rising damp receive, it is very easy to overlook the third and most significant source of dampness: penetrating damp. A survey carried out in the late 1980s, for example, indicated that over 70% of all dry rot outbreaks were attributed to penetrating damp (Bravery, 1991).

In wet, cold climates such as in the UK, penetrating damp in buildings is an ongoing problem. Indeed, in sheer volume terms penetrating damp accounts for by far the largest source of unwanted water that can affect buildings. Its effects therefore cannot be ignored (Kubal, 1993).

Buildings, after all, are essentially climate filters: for people, their belongings, their goods and their accessories. They provide a variety of functions in terms of shelter for protection and privacy, and an enclosure of space, as well as being a climate barrier-modifier. Buildings can also be considered as a 'third skin' (Holdsworth & Sealey, 1992). The first skin is our own skin; our clothes comprise the second skin. The building skin is meant to keep us warm in the cold or cool in the hot as well as dry in the wet.

Unfortunately, because they are relatively crude structures, buildings often do not perform as well as expected. It is not surprising therefore that one of the crucial technical performance indicators of any building is its ability to resist leaks. Next to structural adequacy, water exclusion is probably the most important requirement of any building. Water damage associated with roof or wall leaks can be extensive and very disruptive. The effects of some roof or wall leaks may be obvious, but their source may not be so easily ascertained.

Many newly constructed buildings, regrettably, all too often let in rainwater. It is sometimes very difficult if not impossible to rectify a roof or wall leak immediately. One usually has to wait until the inclement weather has subsided before an effective repair or remedy can be implemented. The consequences of this persistent problem were highlighted by one writer:

> 'Rainwater leaks have, in recent years, given rise to numerous construction delays, disputes and legal actions, with catastrophic financial consequences for contractors, designers and building owners.' (Endean, 1995).

There are two types of penetrating dampness: airborne moisture (rain and snow) and ground-borne moisture. Each of these will be considered in turn. The greater problem of these two categories is undoubtedly airborne moisture penetration. This is not only due to the greater quantities of water involved, but is also attributed to the problem of making buildings in wet climates weatherproof. There is still a dearth of information on how to make modern prefabricated building assemblies watertight. Another influencing factor is that there continues to be a lack of understanding as to exactly how building assemblies leak (Marsh, 1977).

Modern walling components are thinner, lighter and much quicker to install than their traditional counterparts such as stone and brick. The latter rely for their weatherproofing qualities on their thickness and mass to resist rainwater penetration as a result of permeability and poor jointing. That is why the design and construction of modern building enclosures require careful attention to detail, particularly as regards both material absorbency and joint efficiency.

One of the major problems with penetrating damp is that the source of a leak may be in a different place from where it is manifesting itself. Thus tracing a leak can often be difficult and time-consuming. This problem can occur on pitched as well as flat roofs. In both cases the underlying structural framework or services such as pipes and ducts can act as a conduit for a leak that occurs in one place but appears somewhere different (Fig. 8.1).

Effects of penetrating damp

Most types of dampness have similar effects to one another. In particular, however, rainwater penetration, because of the larger volumes of water involved, can cause more extensive damage. The effects are primarily chemical, physical and electrical:

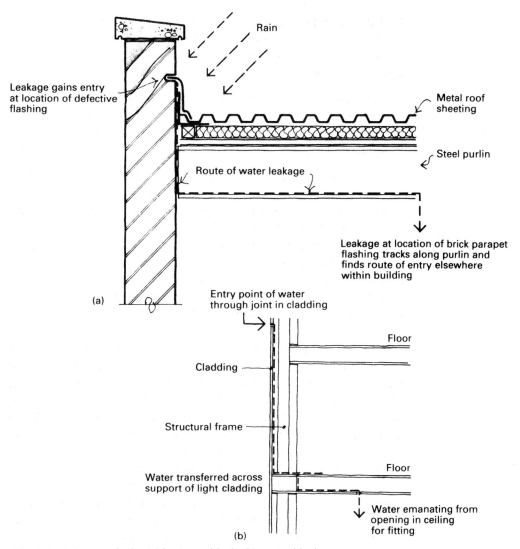

Fig. 8.1 Rainwater leaks: (a) horizontal leak; (b) vertical leak.

❏ chemical: loss of adhesion; sulphate attack resulting in cracking of finishes; efflorescence, rusting and corrosion of unprotected metalwork;

❏ physical: water staining, streaking and damage to finishings; saturation of furniture and soft furnishings; differential moisture movement of finish or background;

❏ electrical: damage to electrical and other equipment; electrical circuits and switchgear may be rendered unsafe.

Water-infiltration-resistant buildings

Requirements

Any well-designed and constructed building must allow for deficiencies in the standard of workmanship. This is where the concept of creative pessimism comes in. One of the fundamental objectives of any new-build project or adaptation scheme is to achieve a water-infiltration-resistant design. According to Botsai (1991):

'... the key to water-infiltration-resistant design is an understanding of the effect of pressure differential on and within buildings. The air pressure difference between the interior and exterior of a building is the identifiable agent responsible for causing all building leaks.

There are several sources of pressure. The obvious are: water head or gravity and wind. The less obvious but equally important are the result of divergent air conditioning pressures in buildings. It must be remembered that these pressures often combine to increase building infiltration problems...

It is critical to understand the distinction between wind load pressure and pressure differential. The two embody divergent principles. The peculiar characteristic of wind load pressure is its capacity to push a water head vertically. In contrast, pressure differential pulls or sucks the water in.'

Pressure differential

It is a fundamental law of building physics in dealing with pressure differentials that all leaks through elements will be carried in the direction of least pressure (Botsai, 1991). Given that external air pressures are usually greater than those inside buildings, water infiltration will continue to be a problem. Thus it is not the quantity of water that is the significant factor in determining the source of the leak, but rather the pressure differential. A case reported by Botsai (1991) on a wall leak highlights this dramatically but informatively. Although the quote is long, and relates to an American building, the lessons it reveals deserve to be considered:

'In considering water-infiltration-resistant design of exterior walls it is important to remember that the force of gravity only becomes critical once the water has entered the building through a flaw. Then, because of the pull of gravity the problem becomes quite complex. The origin of the leak is difficult to find, since water may travel long distances before it is detected.

A classic example of this phenomenon occurred in an eighteen storey building in Texas. Severe water infiltration became evident on the first floor of the building in a total saturation of carpeting. Building occupants sloshed through water while walking over the carpeting. The leak was eventually traced through eighteen floors to its origin on the exterior wall at the ceiling level of the eighteenth floor just below the roof parapet. Once the leak was repaired on the eighteenth floor the first-floor carpeting remained dry.

The important factor in water infiltration of exterior walls above grade is not the quantity of water but the pressure differential. The wall could be fabricated of cloth and saturated with huge quantities of water and if there are no flaws in the cloth and no adverse pressure differential the interior would remain dry. On the other hand, a properly built wall subjected to a slight mist will develop significant leaks when exposed to significant pressure differential. In proof of this, tall buildings are more susceptible to leaks at the upper floors where the pressure differential is the greatest, although as water cascades down the building it increases in volume at the lower levels. It is clear then that wind pressure which causes pressure differentials is the culprit, and not the quantity of water. On a typical wall, as wind pressure increases the potential of pressure differentials increases and in turn the infiltration rate increases.'

Mechanisms of leaks

Pressure differential can cause penetrating damp to enter a building vertically downwards or inclined, or vertically upwards, or horizontally. The first is through rainwater penetration or melting snow, and affects roofs and walls. Vertical penetrating damp may be caused by subterranean pressure, and affects basements. Horizontal penetrating damp affects walls (and sometimes roofs), and may occur through severe wind-driven rain or, more likely, via the subsoil.

The primary pressure differential mechanisms by which unwanted water can enter a building are illustrated in Fig. 8.2 and described below.

Gravity

Rainwater where no wind is present will fall vertically and enter a building if there are any weak points in the watershedding or waterproofing barrier (Fig. 8.2(a)). In such conditions it will hardly wet vertical surfaces such as walls, windows and doors. Any obstructions such as beams or pipes immediately below the source of the leak may cause the penetrating water to travel away from the source and drip in another, completely unrelated, place.

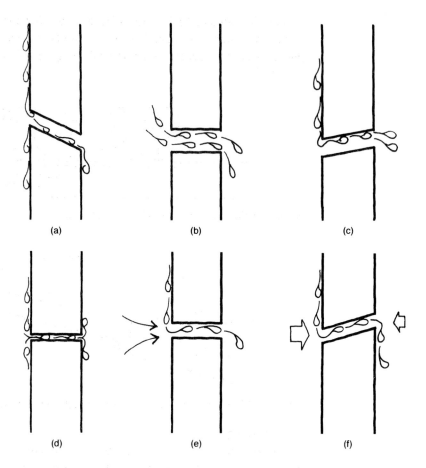

Fig. 8.2   Mechanisms of leaks: (a) gravity; (b) kinetic energy; (c) surface tension; (d) capillary action; (e) air currents; (f) pressure difference.

Air currents
These will cause rain to fall at an angle, and sometimes horizontally (Fig. 8.2(b)). In such cases a building's external walls will become damp. It will also create a pressure difference between the outside and inside of a building, causing draughts, which encourage water to enter the envelope of the building. Storm conditions provide the worst natural damage to buildings.

Wind pumping
In severe windy conditions the moisture may penetrate joints of loosely fixed waterproof sheeting on flat or shallow pitched roofs. The positive and negative pressures exerted on such roof coverings by strong winds can create slight flapping in the sheeting, and this can allow ponded

water to be sucked into the joints (Fig. 8.3). It is for this reason that such a system is not recommended on flat roofs. Leaks caused by wind pumping can also occur at the joints of corrugated pitched roof sheeting (Fig. 8.4).

Fig. 8.3   Wind pumping through a flat roof sheet.

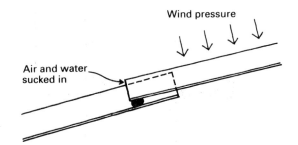

Fig. 8.4   Wind pumping through a profiled roof cladding (Endean, 1995).

Kinetic energy
In severe windy conditions the moisture may penetrate joints of panels by splashing and other forms of kinetic energy (Fig. 8.2(b)).

Hydrostatic pressure
Blocked gutters and outlets will inevitably accumulate excessive levels of water, which will eventually overflow under flashings and into the building. Melted snow will also create this type of pressure (Endean, 1995).

Capillary action
Capillarity creates the surface tension forces that occur between narrow channels, which may cause water or any liquid to flow into or be ejected from the wall surface, even against the effects of gravity (Fig. 8.2(c)).

Surface tension
Water ingress can occur at soffits or the underside of sills that do not have any drip (Fig. 8.2(d)).

Pressure difference
A higher pressure outside than inside may allow water to penetrate through vulnerable gaps in the envelope (Fig. 8.2(f)).

Thermal pumping
Fully supported metal roofing systems may be vulnerable to water penetration by the mechanism known as thermal pumping (Addleson & Rice, 1991). This is caused by the action of wind and rain in summer conditions. It can occur where the joints between the sheets are tightly lapped but not completely airtight, or where rainwater is lying or running against the seam. In hot weather, as rain falls on the warm sheets, the temperature of the roof surface falls, resulting in a partial vacuum in the cavity of the joint. This may be a sufficient force to suck water over the laps of the sheeting joints (Fig. 8.5).

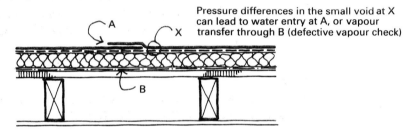

Pressure differences in the small void at X can lead to water entry at A, or vapour transfer through B (defective vapour check)

Fig. 8.5   Thermal pumping.

Contributory factors

Penetrating damp is influenced primarily by both building and human contexts.

The building context

Location and exposure
The position of the building on the site and the degree of protection afforded by surrounding buildings and vegetation will, to a large extent, determine its exposure. North-facing walls, for example, will be more susceptible to the effects of driving rain. A tall building will tend to have more penetrating damp problems in the walls than through the roof, whereas the reverse will be the case with a large-area building. Also, the effect of cooling on moisture absorption and heating on moisture release causes an action known as the 'sponge effect' on walls (Garratt & Nowak, 1991).

Cavity walls

The cavity principle is one of the three main methods of water exclusion (see next item for the other two). The incidence of rainwater penetration through a key element such as a cavity wall depends on a number of factors:

❏ the rain incident on the wall surface;
❏ the leakiness of the outer leaf;
❏ the characteristics of the cavity region, whether open or filled; and
❏ the properties of the inner leaf.

Thickness of walls

Thick porous walls can absorb moisture but still prevent it from penetrating into the building proper. The sponge principle applies here too, and this means that the degree of penetration can be controlled by a sufficiently thick wall (Son & Yuen, 1993). Thin walls need to be relatively impervious to moisture if this problem is to be avoided – the impervious skin principle.

Roofs

The efficient weathering of a roof, whether pitched or flat, is a primary performance criterion. This can be affected by:

❏ the size and profile of the roof areas;
❏ the condition of roof coverings and extent and position of projections;
❏ the efficiency and condition of the rainwater disposal system;
❏ dpc and flashings provision and condition.

Absorbence of the fabric

The more porous the building envelope is, the more it will absorb water. This in itself may not be a problem if the walls are thick enough or are made from aerated material such as 'no fines' concrete. However, depending on the pore size and distribution of the pores, masonry and ordinary concrete materials may be susceptible to frost damage.

State of repair of the fabric

A poorly maintained building will be highly susceptible to rainwater penetration. For example, defective pointing to masonry may cause wind-driven rain to soak its way into the building, gutters and outlets.

Efficiency of inter/intra-element joints

A failure of the joints in an element or of the joints between different components will encourage rainwater penetration. The contrast between raincoat systems and rainscreen systems of wall cladding is important here. Each uses a different approach to keeping the envelope weatherproof, and both are often used as overcladding of existing buildings.

Raincoat systems are relatively jointless (Fig. 8.6). They use a rendered finish on an insulated substrate, and usually contain sealed expansion joints at regular intervals.

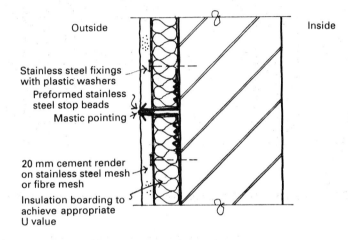

Outside   Inside

Stainless steel fixings
with plastic washers
Preformed stainless
steel stop beads
Mastic pointing

20 mm cement render
on stainless steel mesh
or fibre mesh
Insulation boarding to
achieve appropriate
U value

Fig. 8.6   Raincoat cladding.

Rainscreen systems, on the other hand, can be classified into two main groups: (a) pressure-equalised or (b) drained and back-ventilated, as shown in Fig. 8.7 (Anderson & Gill, 1988).

The human context

Many rainwater leaks can be attributed to either poor design or bad workmanship, or a combination of the two. User misuse and outside abuse should also not be ignored. Some typical examples of all these influences are as follows:

❑ Use of untried or innovative materials can lead to unforeseen incompatibility or durability problems in roof coverings or wall claddings.
❑ Inadequate details: lack of upstand to flashings at abutments and skylights in flat roofs.
❑ Omission of damp-proof courses or seals to parapet walls.

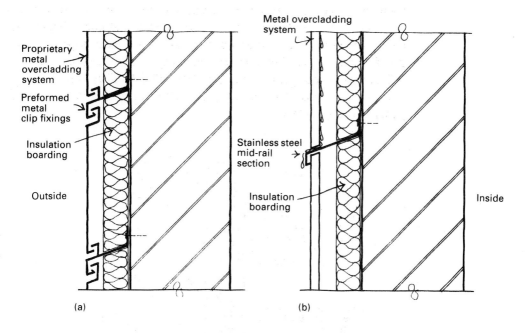

Fig. 8.7   Rainscreen cladding: (a) pressure equalised; (b) drained and back-vented.

❏ Neglect in attending to damage, defects or even routine maintenance work to the external fabric.

❏ Vandalism and abuse of the property by people other than the occupants. For example, thieves stealing lead or other valuable materials from the roof coverings can expose parts of the building to direct rainwater penetration.

❏ Poor workmanship can allow rainwater to enter the fabric. For example, a poorly constructed unfilled cavity wall can provide several routes for water penetration (Fig. 8.8).

Vulnerable locations

The main locations in a typical low-rise building that are vulnerable to penetrating damp are identified in Fig. 8.9.

Penetration through claddings and infill panels

The main types of cladding have been described in Chapter 5. In this section water penetration though claddings and infill panels is con-

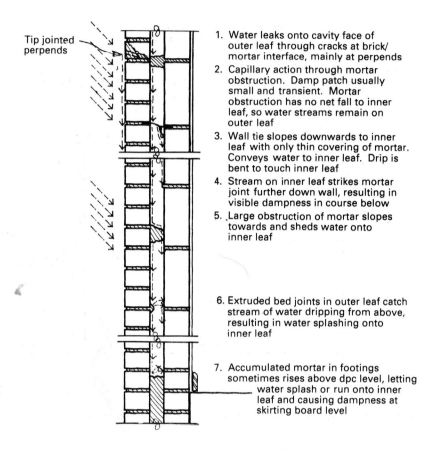

Tip jointed
perpends

1. Water leaks onto cavity face of outer leaf through cracks at brick/mortar interface, mainly at perpends

2. Capillary action through mortar obstruction. Damp patch usually small and transient. Mortar obstruction has no net fall to inner leaf, so water streams remain on outer leaf

3. Wall tie slopes downwards to inner leaf with only thin covering of mortar. Conveys water to inner leaf. Drip is bent to touch inner leaf

4. Stream on inner leaf strikes mortar joint further down wall, resulting in visible dampness in course below

5. Large obstruction of mortar slopes towards and sheds water onto inner leaf

6. Extruded bed joints in outer leaf catch stream of water dripping from above, resulting in water splashing onto inner leaf

7. Accumulated mortar in footings sometimes rises above dpc level, letting water splash or run onto inner leaf and causing dampness at skirting board level

Fig. 8.8   Routes for water penetration across an unfilled cavity (after PSA, 1989).

sidered (Figs 8.10 and 8.11). With claddings the surface is generally 'flush' and uniform except for the presence of joints or openings. Infill panels, on the other hand, are usually bounded on all four sides by a metal or concrete framework. (See Endean (1995) for a thorough study of this problem.) Correct detailing and sealing of the joints with an appropriate (i.e. durable and compatible sealant) is therefore essential (see PSA, 1989).

Penetration through glazing units

One of the most common avoidable failures with glazing, particularly double-glazing replacement systems, is water infiltration. The problem is not so much with the quality of the product, but is rather to do with

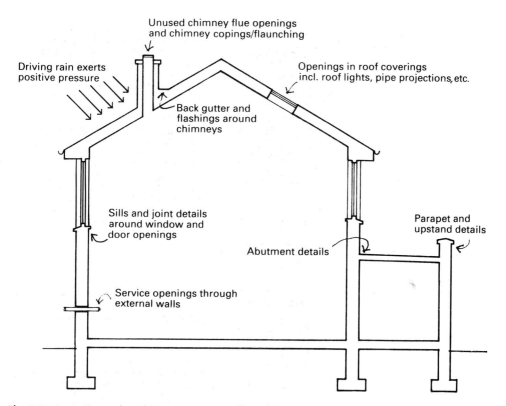

Fig. 8.9 Locations vulnerable to penetrating damp (after BRE Digest 380).

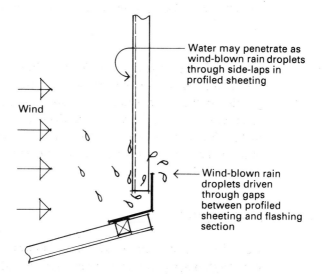

Fig. 8.10 Rainwater penetration through cladding.

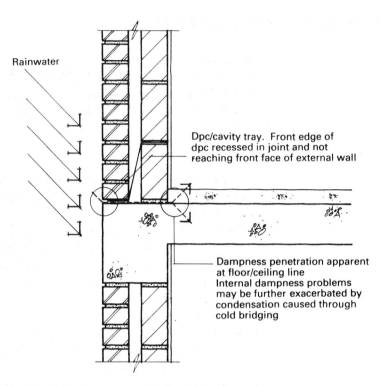

Rainwater

Dpc/cavity tray. Front edge of dpc recessed in joint and not reaching front face of external wall

Dampness penetration apparent at floor/ceiling line Internal dampness problems may be further exacerbated by condensation caused through cold bridging

Fig. 8.11 Rainwater penetration through infill panel.

inadequate installation. The following water penetration problems can be avoided if good practice is observed by the installer:

❏ Poor adhesion of sealant. Dirt, dust, oil and other contaminants will prevent the sealant from adhering properly to the glass or frame.

❏ Incompatibility of sealant. Some non-setting compounds can set, which means that they cannot take wind loading on the glass. This eventually leads to rainwater penetrating the glazing pocket.

❏ Ponded water. This can occur as a result of either a build up of condensed moisture at the sill, or infiltration. Build up of moisture is usually caused by inadequate drainage or blocked drainage and vent holes. Infiltration may have been caused by poorly sealed bead or tape joints. In non-drained or non-vented systems the heel bead can be inadvertently omitted or the tape can have the incorrect compression. This can result in water entering the system but not leaving it. The result is failed units and the decay of frame platforms.

Adaptation influences on penetrating damp

An existing property may be reasonably resistant to penetrating damp if it is well maintained. When a building is adapted or extended, though, penetrating damp could inadvertently be encouraged because of changes to the external fabric. Some of the common pitfalls to be avoided are listed below.

Abutments

Penetrating damp can occur if a cavity tray is not inserted at a vulnerable or exposed abutment. Such trays should always be provided at the flashings of the abutment of a roof to a wall. In these situations the outer leaf of the abutted wall becomes an inner leaf of the wall in the room below. If no cavity tray is provided, excessive rainwater penetration (from, say, persistent wind-driven showers) will saturate the outer leaf and travel downwards unresisted through the skin, which has become an internal wall. This will result in dampness appearing at or near the ceiling below the affected abutment.

It is appreciated, though, that inserting a cavity tray in an existing wall is time consuming and thus expensive. It has to be done in sections not exceeding 900 mm long, and done in staggered fashion to avoid destabilising the outer leaf. Ensuring that the cavity tray is adequately inserted into a convenient mortar joint in the inner leaf can be awkward, but, as has been shown by Endean (1995), it can be done. Still, because of the problems mentioned, many such abutments do not receive a cavity tray.

Roof extensions

Where a pitched roof is extended laterally using the same alignment, the extension roof system should be the same as or compatible with the existing. If, for example, the existing roof profile consists of sarking felt, battens and tiles, it is easy to make the mistake of providing the extension roof with tiles on battens (without counterbattens), on felt, on sarking boards. This will result in the battens being vulnerable to decay should there be any leaks in the roof covering, say as a result of vandalism or storm damage.

Construction joints

Construction or movement joints in flat roof extensions must be provided at the abutment to the existing building. This can be achieved by forming a cavity wall kerb having a minimum 150 mm high upstand and capped with a suitable coping over a dpc.

Surface run-off

New extension roofs should ideally not slope towards the main building, so that any rainwater that may accumulate or not be efficiently discharged does not cause ponding at abutments or saturate any walling. Likewise, the perimeter of a building at ground level should slope away from the main walls.

Protection of structures against ground-borne water

General requirements

When part of a building such as a basement is constructed below ground level, penetration of the soil water would be expected unless measures are taken to prevent penetration taking place. The level of protection required will depend on the basement usage. Categories 1 to 4 are listed in BS 8102, with grade 4 requiring a performance level achieving a totally dry environment (Table 8.1).

While general environmental improvement can reduce the amount of surface water entering a building, the structure can itself be provided with a continuous impervious membrane (type A in BS 8102) that would exclude visible penetration of water as well as providing a vapour seal. (BS 8102: 1990 is an extensive revision of CP 102: 1973.) High-quality concrete can alone provide protection against visible penetration of water, but moisture-vapour transmission may not be wholly excluded (type B in BS 8102).

The extent of water and moisture penetration into below-ground structures may be reduced by modifying site conditions. Surface water should be diverted away from the building on ground levels, preferably paved, that slope away from the building with a 1:40 fall. On sloping sites, a cut-off land drain on the high side should be provided to lead water around the building to the lower level; and a subsoil drain should be incorporated in the stepped construction on sloping ground (Fig. 8.12). Rainwater must be collected and disposed of correctly, away from the building.

The installation of a land drainage system may be effective in lowering the water table (Fig. 8.13). Such land drainage is likely to be more effective in sandy than in clay soil. Field drains should be graded to any or a combination of the following:

❑ soakaways (BRE Digest 365);
❑ open outlets discharging into a watercourse;

Table 8.1 Tanking performance requirements (after CIRIA, 1995a, b).

Grade of basement	Relative humidity	Temperature	Performance level	
			Dampness	Wetness
Grade 1 (basic utility)	> 65% Normal UK external range	Car parks: atmospheric	Visible damp patches may be acceptable	Minor seepage may be acceptable
Grade 2 (better utility)	35–50%	Retail stores: 15°C max. Electrical plant-rooms: 42°C max.	No visible damp patches, construction materials to contain less than air-dry MC	None acceptable
Grade 3 (habitable)	40–60% 55–60% for a restaurant in summer	Offices: 21–25°C Residential: 18–22°C Leisure centres: 18°C spectators 10°C squash 22°C changing rooms 24–29°C swimming pools Restaurants: 18–25°C Kitchens: 29°C max.	None acceptable Active measures to control internal humidity may be necessary	
Grade 4 (special)	50% for art storage > 40% for microfilms and tapes 35% for books	Art storage: 18–22°C Book archives: 13–18°C	Active measures to control internal humidity probably essential	

Note: The limits for a particular basement application should be agreed with the client and defined at the design approval stage.

- ❏ storm water sewer;
- ❏ a sump near (preferably not within) the building, where collected water can be removed by pumping.

In some circumstances, it may be possible to lower external ground levels to below floor levels by excavation and, by using an earth-retaining wall, maintain a low ground level (Fig. 8.14).

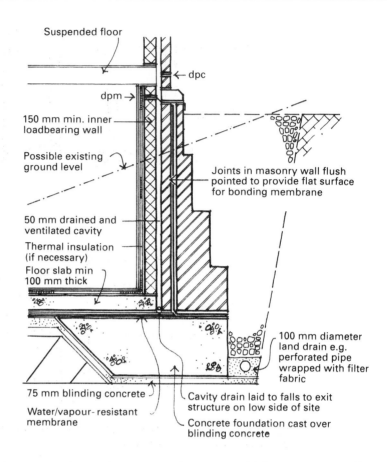

Suspended floor

← dpc

dpm →

150 mm min. inner loadbearing wall

Possible existing ground level

Joints in masonry wall flush pointed to provide flat surface for bonding membrane

50 mm drained and ventilated cavity

Thermal insulation (if necessary)

Floor slab min 100 mm thick

100 mm diameter land drain e.g. perforated pipe wrapped with filter fabric

75 mm blinding concrete

Cavity drain laid to falls to exit structure on low side of site

Water/vapour-resistant membrane

Concrete foundation cast over blinding concrete

Fig. 8.12 Tanking in basement on sloping ground (after CIRIA, 1995a, b).

The design of the structure should be capable of withstanding the pressure of surrounding earth and soil water. The structure should be designed to avoid movement that could damage the impermeable membrane, and particular care is needed in the design of the floor–wall junction to resist earth and water pressure. The functional environmental requirements are indicated in Table 8.1.

Tanking materials

The tanking materials can be categorised into two main groups: traditional and modern. These are summarised in Fig. 8.15.

The impervious membrane should provide a continuous waterproof lining to substructure walls, floors, and foundations. Tanking can be applied either externally (for new build only) or internally (for new build

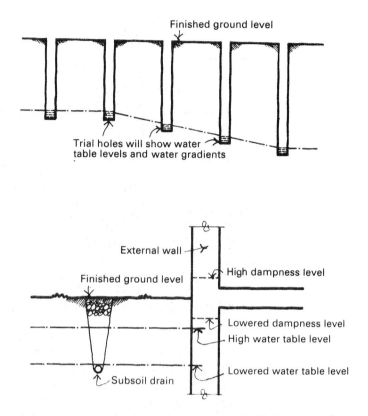

Fig. 8.13 The use of trial holes to determine water table level, and the use of land drains to lower the water table level.

or existing situations). Angle fillets are recommended at the junction of horizontal–vertical membranes, and physical support of these two systems is provided by an additional leaf of brickwork formed in such a way as not to damage, but rather provide protection to, the membrane. Care has to be taken to ensure the continuity of the tanking where special openings using sleeves for pipes or cables are required.

Reinforced and prestressed concrete can provide waterproof structures below ground, while not being fully resistant to vapour transmission. The substructure should be designed to avoid differential settlement, and to avoid cracking. Dense impervious concrete is required, and therefore care has to be taken over the mix design, using a low water:cement ratio, waterproofing admixtures and finer-grade cement, correct placing and compaction. Joints should be constructed with water-bars to achieve full continuity and watertightness.

There are essentially four main techniques used to waterproof substructures (Fig. 8.16):

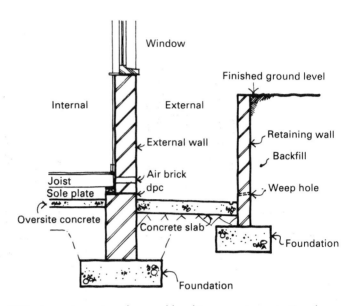

Fig. 8.14   Lowering external ground level to prevent penetrating dampness.

❏ inherent waterproofing;
❏ incorporated waterproofing;
❏ drained cavity construction;
❏ tanking.

In-situ tanking systems

There are three principal types of situation that can arise that may require the installation of tanking systems in existing buildings:

❏ a breakdown of existing masonry or mortar, which provided a relatively dry environment, and the difficulty of repair in an old structure where the original mortar has decayed;
❏ where the outside water table has risen as a result, perhaps, of sewer failure or local flooding, and this has caused water penetration through the previously satisfactory masonry system;
❏ where a change of use of an existing basement is envisaged to provide additional office, living or storage accommodation.

In contemplating internal tanking in these or other situations, it is assumed that the lowering of outside ground levels, either permanently or temporarily, to enable external waterproofing to be applied is not a practical solution.

BS 8102 specifies that a structure should be inherently waterproof and

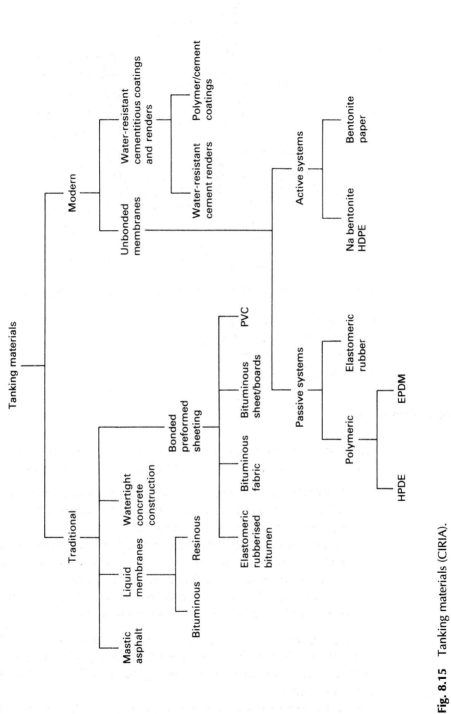

Fig. 8.15 Tanking materials (CIRIA).

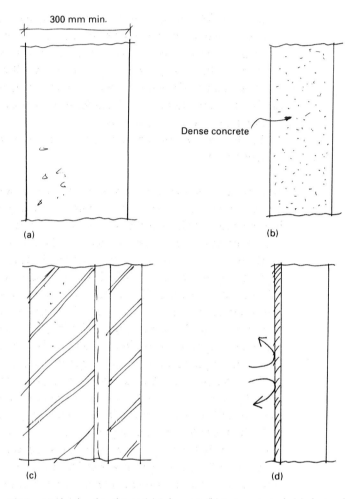

300 mm min.

Dense concrete

(a)

(b)

(c)

(d)

Fig. 8.16   Methods of tanking: (a) inherent; (b) incorporated; (c) drained;
(d) membrane.

has a membrane to achieve this, incorporated together with the structural support for the membrane. Clearly it is difficult for an in-situ-
applied system to perform as well as an externally applied one, but the
Standard provides criteria by which the various systems can be assessed.
It is suggested that the following characteristics apply:

❏ The coating must be impervious to water under hydrostatic pressure.
❏ The coating should be strong enough or have a strong enough bond
   to the substrate to resist hydrostatic pressure.
❏ It should resist the passage of water vapour. If water vapour transmission can occur, condensation could develop in poorly ventilated

areas, which are commonly present in basements. Mould growth or even timber decay could be a consequence, in timber wall lining, for instance. Water vapour transmission may mean water movement through the coating, and the possibility of salt deposition, and could cause failure of the bond between the tanking and the wall.

❑ The tanking should not degrade, lose physical properties or show bonding failure for many years.

❑ Ideally, the system should be able to accommodate some structural movement.

Asphalt, bituminous and polymeric sheeting, cementitious waterproof renders and slurries, chemical coatings and pitch epoxy coatings are used in waterproof tanking systems in such situations.

A three-coat application of mastic asphalt is an effective waterproof treatment for internal as well as external tanking. Although mastic asphalt may be considered old-fashioned and expensive, it is still considered suitable by the BBA (89/2299). Asphalt must be applied to clean dry surfaces, preferably roughened with raked joints. Concrete surfaces may require bush hammering, scabbling, sand blasting or a cement slurry to be applied to provide a good key. Wet substrates must be dried before the application of the first asphalt coat, and a blowlamp can be used for this purpose. A total thickness of 30 mm on horizontal surfaces and 20 mm on vertical surfaces is recommended. It is also important that overlaps of 150 mm are provided at each pour joint.

Asphalt needs physical support, and this can be provided by a masonry wall forming a sandwich construction up to ground level (Chudley, 1995). If asphalt is applied to a wall externally, the ground is dug out, the waterproofing is applied, and then the trench is backfilled with suitable material after the construction of a supporting/protective wall. Complete external tanking installations are usually not suitable for existing buildings because of the break in continuity between the vertical and horizontal membranes. A partial external tanking system can, however, be used to relieve dampness from external walls affected by high water tables. External protection of the asphalt surface is needed to prevent it from being punctured by sharp stones. When asphalt is applied internally, a supporting wall is needed to stop the asphalt from being forced off the main substructure by water pressure. This is usually set back slightly from the asphalt, and a 20 mm space is filled with a lean-mix mortar to keep the asphalt–wall joint solid.

Bitumen sheeting is used in a similar manner to asphalt. The sheet should conform to Class A of BS 743, and for two-layer work a minimum weight of 5.4 kg/m^2 is necessary, with additional strips at internal angles.

A priming coat of bituminous solution or emulsion is first applied to the masonry surface, which should be as smooth, clean and dry as possible. The first layer is bonded with hot bitumen or torch-bonded with laps of 100 mm, and subsequent layers are applied as soon as possible thereafter to clean surfaces.

The use of more rigid corrugated bituminous sheets such as Newtonite lathing combines the advantages of dry-lining and waterproofing. Correctly installed, it can perform well. A moulded polyethylene sheet, Newlath, has been developed, which can be fixed directly to plaster or brickwork using plastics (polypropylene) fixing plugs. As with Newtonite lathing, ventilation can still be provided at ceiling and floor levels. The moulding takes the form of raised studs, which are linked by reinforcing ribs and a polythene mesh is fitted to one side to provide a key for plaster and render. The profile provides channels for air movement. A single layer is sufficient, but laps of at least 100 mm are advised.

A more modern method of internal tanking is to use single-ply polymeric membranes. These are unbonded to the substrate, but must be 'tacked' to it before the protective brick skin is constructed to provide the necessary support. CIRIA (1995a) describes the various other membranes now in use. It should be noted that hydrophilic membranes (such as bentonite composites) are becoming more common for use in external tanking, but they are not suitable for internal tanking (CIRIA, 1995a).

Cementitious tanking methods

There are two main types of cementitious coatings that can be used for waterproofing basement walls: slurries and renders.

When hydrostatic pressure is expected, the cement-based render systems are probably more successful. The most critical aspect of their application is the surface preparation. The masonry or concrete must be cleared of any old covering and the surface hacked or roughened, wire-brushed and washed down. Bush hammering of the concrete surface is necessary to expose the aggregate. Any cracks and crevices are chased out and filled with a quick-setting non-shrinking cement. The first render is usually a strong mix (1 : 3) incorporating a waterproofing and bonding agent, and this is applied as a thrown coat or by brush. This reduces air spaces behind the coating to a minimum. Subsequent coats (1 : 1 : 5 or 1 : 2 : 5) are either trowelled on or splattered to obtain a good key for the next layer. In some systems four coats in all are applied. Service pipes are surrounded by a plug of quick-setting cement. Fixings require special considerations, as it is essential that the render is not penetrated. These products contain quick-setting and usually sulphate-resisting cement, usually with chemical admixtures such as epoxy, styrene butadiene or acrylic resin.

Many types of cement slurry include pore-blocking silicate crystals in their composition. Sodium silicate can diffuse into damp concrete and, by reacting with a free lime in the cement, forms calcium silicate crystals that can grow in size up to 500 μm. Ultimately, it is claimed that they can resist pressures up to 6 bar in 200 mm thick concrete. As well as incorporating sodium silicate, cement slurries may use acrylic and epoxy chemicals, calcium soaps and other waterproofing ingredients. They are normally applied as two-coat seals but, as with cement render, surface preparation is vital for success.

Liquid-applied membranes, particularly the spray-on type, are not normally appropriate for internal use because of health and safety considerations (CIRIA, 1995a). Still, although bitumen emulsions are not suitable for tanking, they can perform well in suspended ground-level floors as a solum treatment (Fig. 5.12).

For a comparison of the performance of the different in-situ tanking systems, see Table 8.2.

Practical tanking considerations

Even the best systems will not perform adequately if they are not applied properly. Generally, considerable skill is needed in tanking work, and in particular a vivid imagination is needed to work out where the water might go and how joints can be effectively sealed. In many situations, cracks and crevices in the wall or floor need to be stopped speedily. Quick-setting cement compositions, based on fondu/Portland cement or Portland chloride mixes, are effective for this purpose. An alternative is a two-pack epoxy putty, which is especially useful in underwater situations, although these may require a period of 24 hours to cure fully.

Other practical considerations include the ease with which the system can be keyed into the rest of the structure, and its repairability in the event of partial failure. Residual solvent odours may also cause problems in the basement areas, where ventilation is restricted.

With some tanking systems it may be difficult to apply a normal plaster finish, because it will not adhere well to a non-porous tanking coating without a mechanical key. This may weaken or puncture the tanking, and the additional load of the applied plaster may also be detrimental. Tanking should always be continued up the wall to a height of 150 mm above outside ground level.

Table 8.2 Comparison of in-situ tanking systems.

System	Criteria				
	Waterproof	Pressure resistant	Vapour proof	Stable	Flexible
3 : 1 sand: cement render	++	++	+	+	No
Render with SBR	++++	++++	+	+	Some
Render with acrylic resin	+++	+++	+	+	Some
Cement slurry coat	+++	+++	No	+	No
Supported mastic asphalt	++++	++++	++++	+++	Some
Pitch polyurethane coating	+++	No (needs reinforcement)	+++	++++	Yes
Two-part pitch epoxy	+++	++++	+++	++++	++++
Newtonite lathing	+++	No	++++	++++	Some
Bitumen felt	+++	No (needs support)			Some
Polymeric felt	++++	No (needs support)	++++	++	Yes
Bituminous emulsions	+	No	No	+	No

++++ indicates that the property is provided to a high degree.

When rising damp is present in solid ground-supported floors in basements, epoxy emulsion coatings applied to the surface may be satisfactory. Polyurethane and polyester resin applications can also work reasonably well. Polythene membranes of at least 1200 gauge laid under a 75 mm thick screed provide the more conventional remedy. However, where hydrostatic pressure is also present, polythene sheeting does not work well. In such cases, a membrane of two-part pitch epoxy liquid coating would be better. Any waterproofing of the floor should always be given to the wall–floor junction as well, which is a major point of weakness.

Remedies for penetrating dampness

General

When penetrating dampness has developed in building elements, various remedial measures need to be considered. The solution adopted will depend on the severity of moisture ingress, the form of construction, and its condition. Isolated or localised leaks, if correctly identified, may pose no major problem. Some water infiltration problems, however, can be extremely difficult to eliminate entirely.

Roofs

If the degree of water penetration through the roof is severe, re-roofing may be the only feasible option. In the case of pitched roofs the existing coverings can be replaced with new or reconditioned material to match the original. Re-slating, with new underfelt and copper nail fixings, is a common example of this practice. In contrast, deficient roof coverings may be overclad with a different material such as profiled metal sheeting. The advantage with the latter option is that the opportunity is provided to upgrade the thermal performance of the roof by including insulation within the overcladding system.

Walls

Repointing
If existing mortar is deteriorating and eroding, then repointing is an obvious solution. It is important to rake out the joints (ideally at least 13 mm) and remove all mortar dust and other deposits from joints before repointing. The mix employed should always be strong enough to be durable but never stronger than the masonry units, and the joint profile should be 'bucket-handled' or 'weathered' (Fig. 5.7). In older buildings lime was probably used in the mortar mix, so this material rather than a straight cement mortar should be used in such instances, otherwise there is a risk that the new pointing will damage the stonework.

Rendering
Straight rendering or rendering incorporating an insulation layer behind

it can often be used to reduce or control penetrating dampness, and this is referred to further in Chapters 2 and 5. Useful guidance is also provided in BRE Good Building Guides 23 and 24.

Slate and tile hanging

Vertical tile and slate hanging can in some situations provide a useful protection for wall surfaces affected by penetrating dampness. The tiles, with moisture-resistant backing such as 'building paper', should be fixed to preservative-treated timber battens using suitably protected nails, copper and steel being the most durable. As the severity of the weather exposure of vertical tile hanging is less than on a sloping roof, the extent of the laps can be reduced to about 40 mm. With slate hanging, the slates should be double nailed at the head and fixed to a 'breather paper' to BS 4016: 1972 or felt backing. Do not use polythene or other vapour check.

This solution, however, is not suitable for wall areas susceptible to mechanical damage, such as that from vandalism or accidental impact.

Weatherboarding

This form of timber cladding, known as ship lap boarding (clap board) or siding in the USA (Scaduto, 1989), was a traditional external finish in certain areas of the UK, such as Kent and Essex. The modern problems of decay in painted softwood used in this way are discussed in Chapter 4. In older buildings, more durable timbers – often hardwoods, such as elm – were used as weatherboarding.

Originally, the boarding would act as the external weatherproof layer fitted to the structural timber studding framework of the wall. It would be either painted or given a weatherproof brush-coating of tar or creosote. Such boarding may eventually succumb to insect attack or local decay, and if general replacement is necessary the boarding should be fitted above a waterproof sarking felt or building paper, using preservative-treated timber.

In recent years, it has become fashionable to use painted or varnished timber, and more recently tongued-and-grooved plastics weatherboarding, as a decorative feature on buildings. Because of the continual need to redecorate such timber, plastics, usually in the form of uPVC, have become more widely used as an alternative.

Like shingles, tile and slate hanging, weatherboarding can be used as a remedy for penetrating dampness where its finish would be regarded as aesthetically acceptable. All these methods may prove expensive, and present problems of detailing.

Silicone treatment

Colourless waterproofing solutions based on 5% silicone or aqueous siliconates have long been used for treatments on porous walls. They can achieve a considerable reduction in rainwater absorption, as well as a reduction in the rate of water evaporation off the treated wall. Their application improves the situation overall, provided that the wall is in reasonable condition and does not contain cracks exceeding 0.15 mm in width. It was initially thought that these protection coatings would remain effective for at least ten years, especially on vertical surfaces. In some cases, however, only a five-year service life can be expected.

If silicone treatments are applied to walls with defects, the water flow may concentrate at the defect points and result in excessive local moisture penetration.

BS 6477 suggests that the new extended-durability testing introduced in 1984 will be met only by a 5% active water-repellent ingredient of either:

❏ organic-solvent-based polymeric silicones; or
❏ organic-solvent-based polyoxoaluminium stearate; or
❏ water-based alkali metal alkyl siliconates.

The performance of water repellents is assessed in this Standard on four types of substrate (Table 8.3), and they are given a group number that indicates whether the product has proved suitable for use on that substrate. Application is usually by brush or spray.

Table 8.3 Classifications of silicone water repellents for masonry (BS 6477: 1992).

Product group	Type of substrate	Example
Group 1	Siliceous	Bricks Sandstone 'Mature' cement mortar
Group 2	Calcareous	Limestone 'Cast stone'
Group 3	Fresh cementitious materials	Repaired, repointed or rerendered surfaces
Group 4	Calcium silicate brickwork	Facing brick

Paint coatings for masonry and cement products

These products combine decorative and waterproofing functions. Sometimes they contain fillers, which can provide a textured finish.

Cementitious paints are usually supplied as a dry powder ready for mixing with water prior to application. They are cement based, and include pigment and a waterproofer. Best application is obtained on slightly damp and porous surfaces. They provide a hard matt and durable surface. They might be considered to have evolved from the limewash or 'distemper', which was used for recoating lime-based external renders on older buildings.

Other paints that have similar properties are emulsified formulations incorporating acrylic resins, bitumens, PVA or styrenes. These systems provide relatively porous coatings so that, while liquid rainwater penetration may be reduced, evaporation of water vapour and drying out can continue, albeit at a reduced rate.

Oil-based paints on external walling, commonly called masonry paints, form impervious films, which have particular moisture-resisting properties. However, they cannot be used on masonry unless it has first dried out. Moreover, masonry paints do not allow the substrate to breathe: thus any residual moisture will be trapped behind the paint film.

Great care is required when applying coatings to masonry. If the purpose is more than purely decorative, and the application is intended to eradicate dampness, it is essential to determine what the actual source of dampness is, and to eradicate that and allow the masonry to dry out prior to the application of the new coating. Any impervious or relatively impervious coating may otherwise accentuate the dampness already present from another source. While it may prove to be a useful remedy for penetrating damp through porous masonry, it could inadvertently make rising dampness or condensation worse.

Substructures

As has already been indicated, basements, cellars and other elements below ground level are susceptible to penetrating damp. This is usually through hydrostatic pressure creating the necessary pressure differential to cause water to infiltrate through the substructures. These elements include floors as well as walls. Leaks in basements can be amongst the most awkward to correct. Sometimes an internal tanking system will be a satisfactory solution if continuity of the membrane can be achieved.

In some cases, however, it is not possible to rectify such leaks because of the lack of access to install an effective tanking system. The following

USA example serves as an interesting case study of resolving a difficult penetrating damp problem (Botsai, 1991):

'... a computer room in the basement of a commercial building developed a small leak through a crack in a wall below grade (i.e. below dpc). The small leak destroyed the humidity control required for proper operation of the computer and forced the computer technicians to work with wet feet on a wet floor, a dangerous condition when dealing with electronic equipment.

... Elmer Botsai first proposed a three-day shutdown to excavate the exterior of the damaged wall and repair the crack properly. This was rejected due to the loss that would be sustained if the facilities were inoperable ($100 000 per day). Accordingly, to solve the problem without a shutdown an interlocking access room on the outside was devised and a small fan used to build up pressure inside the computer room. The small increase in pressure inside the room, retained by means of gasketed access doors, was sufficient to stop the leak. The lesson, from this example, is to think in terms of pressure differentials rather than the volume of water.'

Internal penetrating damp

A common source of water penetration or ingress inside a building is the bathroom or shower-room. Leaks from or around the bath or shower are very common, and their implications are often overlooked (Fig. 8.17). Given the frequency with which people in the West now use showers, this activity will continue to pose a persistent moisture risk inside dwellings and other buildings with such facilities.

Buildings, being predominantly made under non-factory conditions, are not constructed to the same degree of precision as, say, engineering products. It is therefore not always possible to attain a totally watertight seal around either a bath or a shower tray used for showering. Normally the junction between the bath or the shower tray is 'sealed' with a polysulphide mastic. This type of sealant can be very good at resisting water, but requires careful application to be really effective.

The amount of punishment that such joints receive from water under pressure while the shower is in use will test even the best seals. The only way of eliminating this problem is to use a shower cubicle with integral tray and walls, or a bath that has raised sides at its junction with the walls. Such accessories, of course, may be expensive.

A simple and effective solution for existing situations in dwellings can obviate the need for these expensive cubicles. This involves installing

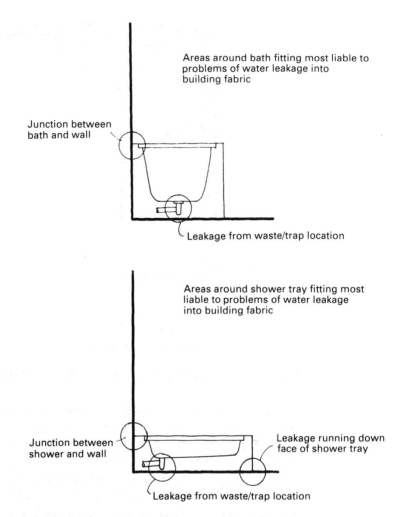

Fig. 8.17   Problems of water leakage from baths and showers.

shower curtains around the sides of the bath-shower or shower cubicle as well as at the front (Fig. 8.18). In most cases the curtain is provided only to close the 'compartment' in the bath-shower or shower recess.

Shower-rooms and baths in non-residential buildings are usually more extensive in size and number than those in dwellings. Correct design and specification of floor, wall and ceiling finishes, as well as adequate heating and ventilation, are vital if condensation and penetrating dampness problems are to be minimised in such facilities. The revisers have encountered several instances, for example, where absorbent finishes such as mineral fibre 'acoustic' panels have been used for suspended ceiling tiles in shower-room areas! Not surprisingly the tiles at the

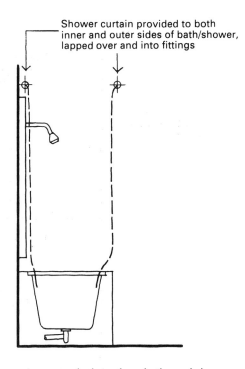

Fig. 8.18   Preventing water leakage from baths and showers.

shower-head positions were badly warped and stained. In cases where the shower-room ceiling is a concrete slab soffit, numerous specks of mould growth are not uncommon sights because of the condensation problems associated with these facilities.

Dampness problems associated with commercial shower-rooms could be overcome if they were able to be installed as modular facilities built to factory standards. Problems in existing as well as new non-residential buildings can be minimised by adopting the following measures:

❑ adequate overhead mechanical extract ventilation providing the minimum number of air change per hour to suit the conditions;
❑ using large-size panelling for wall and floor finishes to keep the number of joints requiring sealing to a minimum (the more such joints, the greater the potential for joint failure to occur);
❑ internal insulation behind wall and ceiling finishes to reduce surface condensation and mould growth;
❑ avoiding sharp junctions at the joints between floor–wall and wall–ceiling;
❑ drainage channels that drain away from the walls and towards the centre of the floor.

Risk of rain penetration

See Table 8.4.

Table 8.4   Risk of rain penetration with cavity insulation (after Haverstock, 1988).

	← Reduced risk		Increased risk →	
Insulation	Partial fill with 50 mm residual air space	Filled with expanded polystyrene beads or mineral wool	Partial fill with less than 50 mm air space	Filled with UF foam
Cavity	Greater than 50 mm		50 mm or less	
Pointing	Bucket-handle or weathered tooled	Flush	Recessed tooled	Recessed untooled
Mix	Cement–lime–sand		Cement–lime–sand or masonry cement: sand	
Finish	Cladding	Rendering		Paint

Tracing penetrating dampness

In many cases the source of a penetrating dampness problem is obvious: a missing or defective roof covering, a broken or open wall panel/joint, to name but two examples (Newman, 1988). Other sources of penetrating damp, however, are not so easily traced. Some leaks may be concealed behind apparently dry linings or may be diverted from the entry point by the construction or services. Some penetrating damp may run horizontally before appearing (Fig. 8.1(a)); others may first drip vertically before travelling laterally (Fig. 8.1(b)).

It takes time, effort and a sound knowledge of building construction to be able to ascertain the actual source and cause of these more complex leaks. Where penetrating moisture has been identified as the cause of the dampness, but its source has not been pinpointed, a number of steps need to be undertaken (see Chapter 9).

The tracing of leaks can be aided by the following:

❏ inspecting the property in detail during periods of heavy rainfall to note water entry points;
❏ using localised water testing with a kettle container at suspect positions;

❏ applying dyed water at suspected leak locations to distinguish the various sources of moisture;
❏ dismantling or removing suspended ceiling tiles or other panels that may be hiding the location of the leak;
❏ studying the 'as built' drawings to identify likely moisture paths.

Avoidance measures

Penetrating damp can be minimised if essentially the pressure differential plane is properly located and has adequate continuity. Ideally, the pressure differential plane should in most cases be positioned as close to the exterior of the building as possible. Thus if moisture does penetrate the extreme outer surface and then contacts the major differential plane it is easier to remove (Botsai, 1991). See Verhoef (1988) for recommendations for the design of building facades to minimise problems such as 'rainshadowing', seismic staining, and general soiling.

As regards the continuity of the pressure differential plane, the following measures are normally needed.

Watershedding
The building should have adequate watershedding and draining abilities to discharge any rainwater efficiently. Gutters and flat roofs should have falls not less than 1 in 40.

Weepholes in parapet walls are best provided above the flashing and cavity tray. This will allow any water entering the cavity to drain away to the nearest outlet in either the parapet gutter or roof. The common alternative is to allow such water to escape from the weepholes on the outer facing side of the parapet. The problem with the latter solution is that any leaks from the weepholes will probably cause seismic staining on the external wall, which is not only unsightly but may also encourage frost attack in the wetted bricks. Overflows, pipes or outlets should be incorporated in parapet gutters, to avoid any build-up of water created by a blockage or defect in the rainwater goods. A discharge from an overflow would serve as a useful indicator of a problem as well as relieving any build-up of rainwater.

Waterproofing
Roof surfaces with minimal inclines, such as valley and parapet gutters and flat roof areas, should be made totally watertight. In new-build cases, water tests should be carried out at critical locations to ascertain the effectiveness of the seal. Such tests are usually not recommended in

existing situations because, like water tests in old drains, they could cause more problems than they solve by overloading a possibly inherently weak but watertight lining.

Damp-proof detailing
Detailing of damp-proof courses and membranes should adopt the quality and form suggested in authoritative and experienced guides, such as BS 8215 and Duell and Lawson (1983) respectively.

Cavity trays and dpc's
Abutments between roofs and walls, lintels over openings, and jambs and sills of windows should have provision for cavity trays and dpc's to prevent water penetration. These should contain weepholes at every 900 mm centres horizontally.

Flashings
Upstands and skirtings at abutments, rooflights and other projections should have minimum 150 mm high flashings properly raggled into the walling. The flashing should be inserted below (not above) the wall damp-proof course/cavity tray; otherwise, penetrating damp will occur at the top and behind the flashing.

Weatherings
Adequate ovehangs, copings, drips and throats should be provided at all projections and exposed cappings.

Pointing
Closed joints to cladding panels and the surrounds of openings should be sealed with a high-performance polysulphide mastic.

Windows
All new and replacement windows must have provision for background ventilation in the form of trickle vents in the head of the frame. Such background ventilation should be at least 4000 mm^2. In addition, weepholes should be provided in hollow frames to allow any condensed or penetrated moisture to escape to the outside. The window surrounds should be adequately sealed against driving rainwater (PSA, 1989).

Doors
The surrounds to doors and windows are very vulnerable to damp penetration. Careful consideration therefore must be made to the provision of dpc's at jambs and heads, and all surrounds should be pointed

with a high-performance mastic. Weatherbars should be provided at all external door thresholds (PSA, 1989). Ideally, the outside ground level should slope away from the building, particularly from any doorways.

Where doorways open onto flat roofs or areas outside a building prone to flooding, it is essential to form a kerb upstand at the threshold, and this should be not less than 150 mm high.

Maintenance
Gutters and outlets should be cleaned out regularly (i.e. every six months) during the autumn and spring.

Chapter 9
Diagnosis and Therapy

Methodology

Diagnostic criteria

The successful investigation of any building problem involves a careful, systematic and rigorous approach. Any such investigation must be undertaken with an open mind and in a clear methodical way if mistakes or misdiagnoses are to be avoided or at least minimised (Addleson, 1992). It is vital not to be biased or become prejudiced about the cause of apparent building faults. Vested interests and the need to achieve a sale of a specific remedy may cloud an investigator's diagnosis.

Moreover, investigating such problems is often fraught with difficulties and red herrings. It is not possible to dismantle an existing building completely to see the locus and extent of a defect. Thus direct accessibility may be impossible or awkward. Moreover, there can be more than one possible cause of a defect, especially one involving dampness (Ransom, 1987).

Investigating building defects is therefore a demanding and difficult exercise. It often brings relatively little financial reward and yet places a heavy professional burden on the surveyor. This is the worst of both worlds for any building investigator. Dampness investigation is no exception. The fee that can be expected for such a service may amount to only a few hundred pounds. The costs of incorrectly diagnosing a dampness problem can run into thousands of pounds in repairs and legal damages.

Getting the diagnosis of dampness right is crucial if both an inappropriate remedy and an ensuing claim for negligence are to be avoided. For example, mistaking rising damp for condensation (or vice versa) and implementing the repairs (wrongly) considered necessary will not resolve the problem (Fig. 9.7). The remedies for rising damp are quite different from those required for tackling condensation. Thus any repairs intended to resolve an apparent rising damp problem will be a waste of time and money, as they will have little or no effect (Howell, 1994).

Dampness is a natural condition in our environment; indeed, excessive dryness can be as uncomfortable to the occupants and as undesirable to many materials as excessive dampness. With several of the modern diagnostic methods now available, relatively low levels of dampness can in fact be detected. What we need to ascertain is the presence of excessive or significant dampness, and so all data must be interpreted with that in mind.

Prognosis

The recognition of the implications of dampness in a building and its effects on building elements and components is an essential aspect of correct diagnosis. The primary objective is to determine the source of dampness so that appropriate remedial measures can be taken.

It is thus vital that consideration be given to a prognosis of the dampness. This means considering not only the effects of dampness but also the consequences of not dealing with the problem. For example, neglecting a roof leak could lead to fungal attack in timbers within or near the vicinity of the dampness. The surveyor must point out these implications to the client if a claim for negligence is to be avoided.

Inspection and appraisal

General

When undertaking an inspection of a building, it is always wise to adopt a standard routine. The same methodology applies when investigating a dampness problem. A basic two-stage approach is usually the most effective for general inspection purposes:

❑ stage 1 – desktop study;
❑ stage 2 – on-site survey.

Stage 1: Desktop study

This phase involves an examination of documentation such as:

❑ as-built drawings (if available);
❑ the original specification;
❑ maintenance manual (if produced);

❏ relevant correspondence;
❏ reports on past problems and repairs.

It also includes the surveyor's liaising with the occupants/owners for direct feedback on the nature, timing and extent of the problem.

A thorough desktop study will enable the surveyor to be armed with information that may shed light on the history of the dampness problem being investigated.

Stage 2: On-site survey

This stage can be broken down into the following.

Reconnaissance
The surveyor should undertake an overview of site and property. This will indicate the context of the property, and may shed light on common building problems in the area.

External inspection
Not only does this provide the opportunity of noting materials and methods of construction, but also obvious signs of water penetration can be recorded. For example, the presence of organic growths such as algae, lichen and moss are evidence of dampness (BRE Digest 370). A sketch plan of the ground-floor area is also helpful as a means of 'plotting' the signs, and serves as an aid when the internal part of the inspection is undertaken. A checklist such as that given in Appendix F is sometimes helpful in ensuring that all aspects are considered during the survey and are subsequently recorded in the report.

Internal inspection
A detailed examination of affected parts of the interior must be carried out. Again, a checklist such as the one shown in Appendix G can provide a useful guide when assessing the dampness.

External inspection review
It is important to review the exterior fabric to check for faults or symptoms thath may have been overlooked earlier, and to obtain confirmation of any initial diagnosis.

Dampness investigation

In an investigation into a dampness problem, the seven-step procedure suggested by Garratt & Nowak (1991), should follow the foregoing two-phase approach:

(1) Visit the site. Is there a dampness problem?
(2) Diagnose the source of dampness (see Fig. 9.2).
(3) Look at the factors: occupants, building, weather.
(4) Identify the cause: too cold, too wet, poor ventilation.
(5) Select the remedy: if condensation, heating, ventilation, insulation.
(6) Apply the remedy.
(7) Follow up. Monitor effectiveness of therapy.

General survey equipment

For the normal survey for dampness and associated problems, the surveyor needs certain basic items of equipment, including:

❏ a torch and a mirror (on a long handle with swivel end);
❏ tools for lifting floorboards (e.g. hammer, feather-splitting chisel, and saw);
❏ a moisture meter and spare batteries (see below, page 264);
❏ hand lens (e.g. magnifying glass);
❏ tape measure (an ultrasonic type is quite useful here);
❏ timber probe;
❏ folding ladder.

For a more detailed survey other items may be useful, such as:

❏ a metal detector;
❏ a camera and binoculars;
❏ a digital hygrometer;
❏ an optical probe such as a borescope;
❏ a calcium carbide meter (Fig. 9.1(a));
❏ an incremental borer.

In many instances, however, especially when a dispute is in prospect, there is no substitute for the removal of samples for laboratory analysis. The most widely adopted method of taking samples of masonry is to take drillings with an electric drill using an 8–10 mm diameter bit. Samples of timber affected by fungi can be taken to a laboratory for cultivation to confirm or determine the species of rot.

For special purposes, other more sophisticated and specialised techniques have been adopted. X-ray and gamma-ray sources can be used to record hidden internal details and inconsistencies in porosity. Inclusions and discontinuities can be demonstrated by radiography. The use of

(a)

(b)

(c)

Fig. 9.1   Typical moisture meters: (a) 'Speedy' carbide meter (courtesy of Protimeter plc);
(b) 'DampCheck' meter (courtesy of Protimeter plc); (c) Digital hygrometer (courtesy of Protimeter plc).

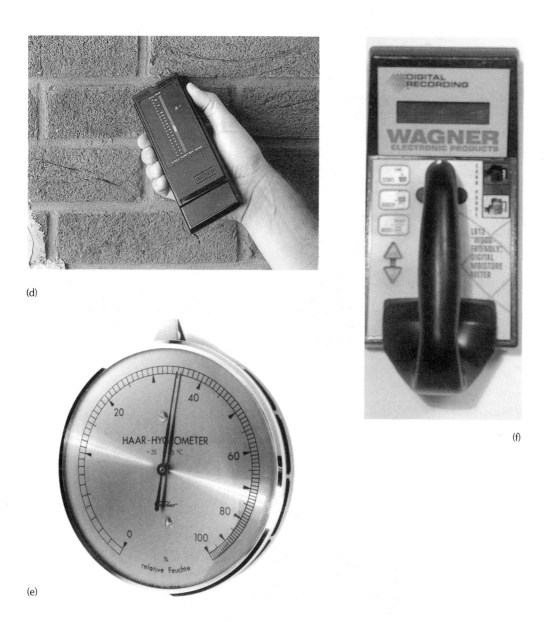

(d)

(f)

(e)

Fig. 9.1(cont'd)   Typical moisture meters: (d) 'Mini C' pin-type moisture meter (courtesy of Protimeter plc); (e) Hair hygrometer (courtesy of Vega Instruments); (f) 'Electromagnetic' wood moisture meter (courtesy of Wagner Europe Ltd).

Polaroid film with X-rays, with on-site processing of the film, facilitates non-destructive examination of inaccessible components. Gamma rays are required for examination of more dense materials such as stone or concrete, and specialist surveys with these techniques are now available. The detection of infrared wave components of heat energy enables identification of 'cold bridges' to be carried out, and thermal image photography can be used to detect hidden components (BRE, 1991). Portable ultrasonic testers can be used to examine welds, and are sometimes used for the examination of steel and concrete reinforcement. See Hollis (1991) for a comprehensive assessment of these more sophisticated items of building surveying equipment.

Moisture detection

Methods

There are basically three methods of determining and/or measuring moisture in building materials, as listed in Table 9.1. However, before looking at these three groups in more detail, we need to consider the significance of measuring the moisture content of building materials.

Table 9.1 Methods of moisture determination.

Method	Type	Uses
Gravimetric	Oven-drying method	Masonry materials mainly
Chemical	Carbide method	Masonry materials only
Electrical	Conductivity Capacitance Radio frequency Electromagnetic	Timber; but can be used with care on masonry materials Timber only

Significance

It has been emphasised by experienced writers in the field that 'dampness cannot be measured by moisture content in the range of materials met in buildings' (Oxley & Gobert, 1994). Thus percentage moisture readings in building materials other than wood are not very meaningful. They can, however, be helpful in determining whether a material is wet or dry, and can give an indication as to where to look further for the cause of the dampness. But taken at face value they can be very misleading, for the effects of electrolytes (e.g. salts and certain wood preservatives such as CCA or boron) can exaggerate readings. Furthermore, different masonry materials have different 'dry' and 'wet' conditions (Table 9.3).

Measurement of relative humidity

When inspecting a building for dampness, it can usually be helpful to determine the relative humidifies (RH) of the internal and external environments. A knowledge of the humidity and the temperature of the building fabric will confirm whether condensation is occurring. It should always be remembered, however, that humidity can change very rapidly, not only from day to day but even from hour to hour. So the fact that the humidity is not high at the time of a survey does not mean that condensation will not occur in a few hours' time. That is why electronic condensation tell-tale devices such as the Protimeter Damp Check (Fig. 9.1(b)) can be so helpful.

Relative humidity can be measured by using one of three instruments: a wet and dry bulb whirling hygrometer, a hair hygrometer (Fig. 9.1(e)), or an electronic hygrometer. Whirling hygrometer meters have a thermometer that can be used as a wet bulb if the surface is covered with filter paper moistened with distilled water. The interpretation of the readings to determine the dew-point temperature can be obtained from a psychrometric chart (Fig. 6.1). If using a hair hygrometer an instant dial reading is obtained. A digital read-out can be obtained when using an electronic instrument such as the Protimeter Diagnostic Hygrometer (Fig. 9.1(c)).

Gravimetric (oven-drying) method

The most accurate method of determining the moisture content of a material is to take a sample, weigh it, dry it to constant weight in an oven at a suitable temperature – usually about 100 °C – and then reweigh. The dampness is expressed by the weight loss achieved by this drying as a percentage of the oven-dry weight of the material being examined.

This method is relatively time-consuming, and is only appropriate for special investigations. It does, however, have the merit of being easily adapted to determine both hygroscopic salt and capillary moisture in a rising damp situation. This technique (see Appendix E) is of special use when the performance of a damp-proof course or plaster is called into question (BRE Digest 245).

The electrical and chemical methods of measuring the moisture content of building materials are considered in the following sections.

Moisture meters for building materials

Background

No doubt the most widely used hand-held instrument for the diagnosis of dampness in building materials is the electrical conductance-type moisture meter. Correctly used it can prove to be a helpful tool for assisting diagnosis. Dampness in building materials cannot be detected by human senses below 85% humidity equilibrium. For instance, timber does not feel damp to the touch below 30% moisture content, which is about 97% RH (Oxley & Gobert, 1994). Hand-held moisture meters can thus help the surveyor to detect dampness that may not be obvious at first.

But, as with all such instruments, the results must be interpreted in the light of all the relevant data obtained from the survey. Moreover, recent studies have questioned the use of such meters on masonry materials (Howell, 1995). They are after all mainly intended and calibrated for use on timber, not masonry (see next section below). Accordingly, extreme care must be exercised by surveyors when interpreting moisture meter readings on non-wood materials (see also Appendix M).

Moisture meters for wood

There are now four main types of hand-held wood moisture meters used by surveyors: contact, pin, electromagnetic-wave, and radio frequency emission instruments. The first two are the conventional types of moisture meter; the other two are the more modern types.

Electrical moisture meters

The electrical conductivity/resistance of most building materials varies with moisture content. Therefore instruments that measure conductivity offer a practical on-site technique for determining whether a material is damp. It should always be appreciated, however, that these instruments do not actually measure moisture content; they measure conductivity. Thus readings using this type of meter, for example, on walls with foil-backed paper with certain types of coating, or with carbon granules, will produce excessively high readings, which have nothing to do with dampness. In addition, because such meters are not calibrated for use on masonry, readings on these materials massively overestimate the amount of moisture present (Howell, 1995). Previous studies (Howell, 1994) have suggested that not many surveyors are aware of this problem, and believe that the meter readings do, in fact, give a true indication of moisture content.

Two forms of electrical moisture meter are currently in use: the conductivity (i.e. pin-type) and the capacitance (i.e. contact-type) instruments. Older versions of both are of the analogue form, with a dial that shows WME readings on a scale, highlighting three key zones: dry, normal and wet. Some of the more modern ones are electronic, which give a digital read-out.

Conductivity meters

These instruments use needle probes, which are placed on or (if it is soft enough) in the material to be examined. The electrical resistance of the material between the probes is measured and recorded on a meter scale. It gives readings in terms of wood moisture equivalent (WME). When examining masonry it is usual to note such readings on the arbitrary WME scale, because actual moisture content percentages suggest a standard of accuracy that is not really warranted.

Capacitance meters

Conductivity plates are placed on the surface of the material. The readings obtained measure the fringe capacitance of the sensor, which is influenced by the moisture content. These instruments have the advantage that the surface being examined is not marked by needle probes.

Protimeter plc is one of the main suppliers of both pin and contact meters, as well as a whole host of other types of moisture detecting equipment.

Meter readings in depth

Where readings in depth are required, there is a range of deep probes available for use with electrical moisture meters. A hammer electrode with insulated needles can be driven into softer materials and a WME gradient determined, possibly up to 30 mm from the surface of the sample. Pre-drilling into masonry or timber enables longer probes to be inserted, perhaps up to 250 mm with very deep probes.

If high WME readings are obtained throughout the depth of a sample, this usually suggests rising damp rather than condensation or penetrating damp. Condensation usually only gives high readings at the surface. Penetrating damp may give high readings throughout the depth of the wall, but usually they will be higher towards the source and often lower towards the surface.

Effects of salts

Apart from moisture, the conductivity of materials can be affected by the presence of salts, which are almost invariably present in rising damp and sometimes found in penetrating damp situations. The presence of these salts will increase conductivity, especially at the edges of damp areas, where drying out and accumulations of salts occur. Many of these salts are likely to be hygroscopic and further add to the moisture present, especially during periods of high humidity. Such salts are deliquescent, and this can often exacerbate a dampness problem.

For these reasons, meters will not only measure moisture present, but also salt accumulations. Readings will be similar, for different reasons. Thus it is only with a knowledge of the way in which moisture moves in buildings, and the reactions of building materials to moisture, that correct interpretation can be achieved.

Electromagnetic wave moisture meters

These more modern meters are designed solely for measuring moisture in timber. Thus they should not be used on masonry materials. Electromagnetic moisture meters generate a three-dimensional field that penetrates through the surface of the wood, and they are relatively unaffected by the the temperature of the timber or the surrounding environment. They were originally developed in the USA for use in the forestry and timber industries, but are now marketed in the building sector as well.

According to one manufacturer of this instrument (see below) the method is non-invasive, and works by measuring the density of the timber. As the timber gets wetter so does its density. An adjustment is made to the meter reading depending on the species being measured, and this adjustment is shown in the extensive tables that are provided with each meter. The meter itself is not re-calibrated; the operator makes an adjustment to the reading. Digital recording versions of the instrument are now available alongside the conventional dial-type.

Wagner Europe Ltd is a major supplier of this type of instrument.

Moisture meters for non-wood materials

Electrical conductance meters

For the reasons indicated above, surveyors should use these instruments with great care when applying them to non-wood materials such as brick, stone and plaster. First of all, they are designed primarily for use with timber, and are calibrated accordingly. Second, they are subject to many environmental variables that can dramatically affect moisture readings.

For example, the temperature of the timber and the chemicals trapped within the wood can distort the readings. Such readings should always be corrected for any difference in temperature above 25 °C.

The resultant WME value from an electrical-conductance meter when used on masonry is in itself not very meaningful. For instance, in tests carried out on damp bricks by Howell (1995):

'The conductance meters tested gave readings up to 100, which may lead surveyors to record a reading of 100% moisture in material which is capable of holding 4% moisture. At lower moisture levels, the meters give readings in the range of 25 to 40 for actual moisture contents between 0.4% and 1.6%. If readings are taken at face value, then, the result is to massively overestimate the amount of moisture present...

... However, it must be remembered that if a meter was produced which could indicate the moisture content of masonry materials, it would still be limited in its applications. Knowing the moisture content of a material is only meaningful if you know its original, dry, porosity. Furthemore, knowledge of percentage moisture contents does not indicate the source of that moisture.'

Radio frequency meters

A recent development by Protimeter plc is a moisture detector using a radio frequency emissions transmitter housed in the back of a modified pin-type moisture meter. The advantage of this facility is that it provides a non-destructive assessment of below-surface dampness.

Carbide meters

This type of moisture meter was originally issued under the trade name Speedy. Protimeter plc is now the main supplier of this instrument in the UK. When using a calcium carbide moisture meter, a sample of masonry is first obtained (e.g. enough crushed stone, brick, mortar, etc. to fill a small matchbox), usually by drilling the wall, mortar or plaster. The sample is weighed with a simple balance on the meter, and is then reacted with a small measure of calcium carbide in a gas pressure vessel. The resulting gas pressure generated can be measured on the gauge at the base of the flask (Fig. 9.1(a)). The quantity of gas generated is directly proportional to the moisture content of the sample. Thus calcium carbide meters have an apparent advantage in that they measure the actual moisture content of the sample rather than WME. However, they do not automatically demonstrate the source or characteristics of that moisture (i.e. is it hygroscopic or capillary in origin?). Special procedures have to be followed, involving conditioning the samples, before these questions can be answered (see Appendix E).

While the actual percentage figures of moisture are helpful, alone they will not differentiate between different forms of dampness. Most building materials are naturally variable in their moisture capacity and relationships, and moisture profiles have to be established before a correct diagnosis can be arrived at.

Moisture content of building materials

As has already been stated, building materials react to penetrating moisture and atmospheric humidity in different ways. In order to try to interpret moisture content of materials and to decide whether moisture levels are excessive, we need to know the typical level of moisture of the material under acceptable conditions.

Moisture content is usually expressed as a percentage of the oven-dry weight, but as the basic density of a material varies, so does the actual moisture present in that material when expressed as a percentage. Thus at 2% moisture content, a cubic metre of limestone whose density is $2120 \, kg/m^3$ would contain 42.4 kg of water. At 2% moisture content, a cubic metre of pine with a density of $530 \, kg/m^3$ would contain 10.6 kg of water. However, the capacity of the material to absorb and hold water will be related to porosity. An open porosity is quoted in Table 9.2. This is the volume of pore space accessible from the exterior, expressed as a percentage of the total volume. Surface and interconnected internal pores are included, but internal voids that are sealed off are not. Generally, porosity decreases with denser materials. Thus we can expect materials of lower density and higher porosity to have a greater capacity for 'moisture content'. Maximum moisture content percentages are

Table 9.2 Density and porosity of some building materials.

Material	Density (kg/m^3)	Porosity (%)
Limestone and sandstone	2120	24
Brick (clay)	1740	26
Pine	530	43[a]
Fibreboard	87	44

[a] Total void volume at 12% moisture content is 63%.

much lower for dense, less porous materials (e.g. 2% in granite might be above total saturation, whereas in wood it would be scarcely detectable).

Typical moisture content percentages for common building materials could be proposed for a dry and a damp state, and this should be considered when interpreting random readings of masonry moisture contents (Table 9.3).

Table 9.3   Possible actual percentage moisture contents for some common building materials under 'wet' and 'dry' conditions.

Material	Dry	Wet
Plaster	Below 1	>1
Bricks	1–3	>3
Cement mortar	1–2	>2
Lime mortar	2–5	>5
Wood	8–16	>20

The importance of measuring the actual moisture content of damp masonry has received some prominence, as in both BRE Digest 245 and BS 6576 it is stated that a wall with a moisture content at the base of less than 5% is unlikely to have severe rising damp. Thus one reading of below 5% with a carbide meter taken near floor level could give the impression that no significant rising damp was present. In contrast, the establishment of a moisture profile (see Table 9.5) would show the origin and vertical distribution of the dampness, and would lead to a more accurate and valuable diagnosis.

The actual moisture content of most building materials may also depend on the amount of water deriving from adjacent penetrating dampness, as well as from hygroscopic salt action. These salts might be present in the building materials initially, but their influence is on the whole relatively minor compared with the 'contaminating' salts that result from rising dampness.

Wood and other cellulosic materials

In wood and other cellulose-based materials, a different situation arises. This is because of the special attraction and hygroscopicity of cellulose to water and moisture vapour. Wood and other cellulose-based materials contain cells that have a considerable void area (perhaps as much as 60%), and so the potential of moisture capacity is much greater than for any other structural building material (see Chapter 4). In fact, the hygroscopic moisture content (equilibrium moisture content) is closely related to ambient humidity, so generally much higher moisture contents would be expected in timber than in almost any other building material.

The conductivity of wood as measured by a moisture meter is quite closely related to the moisture content over the 5 to 25/30% range, and with an accuracy of $\pm 2\%$ the meter provides quite accurate values. Of course, moisture gradients can exist in wood, and salts can also be present, which could lead to additional increases in conductivity and apparently higher moisture contents being recorded.

Water-borne salt-type preservative treatments can affect the moisture readings on timber. Tests have shown that at the lower moisture range, up to about 15%, the effect is small, but as moisture levels rise above 18%, much increased conductivity is obtained, and falsely high readings result.

Solvent-based preservatives seem to cause little practical effect on the meter readings.

Moisture patterns

As has been described, in many situations in buildings where dampness of whatever source is suspected, moisture contents (apart from readings in wood) obtained by the different methods available do not always aid diagnosis. Indeed, they may offer conflicting evidence.

The only practical measure is to establish moisture patterns across the affected area: for example, up the full height of the wall. This method assumes that the area under inspection was constructed of materials generally of similar constitution (e.g. sand, lime, cement, bricks) and that any detectable variations can be associated with the excess dampness being investigated. When these are considered alongside the other visual signs of dampness, a reliable diagnosis can usually be obtained.

Typical readings obtained from such investigations show how the individual components of dampness can be assessed (Table 9.4).

Surveyors should of course be wary of interpreting moisture meter readings in isolation (see Appendix M). As ever, the aim is to locate the source of the moisture once its significance is ascertained. Moisture patterns are always established when carrying out an examination to determine hygroscopic and capillary moisture content, which involves taking a series of samples from up a wall to above the maximum height of rising damp (see Appendix E). In addition, a salts analysis of a sample of plaster near the upper end of the damp patch will be another aid for obtaining a correct diagnosis (see Appendix D).

Table 9.4   Typical moisture patterns on an external wall using a pin-type moisture meter.

Height above ground floor (m)	WME meter readings (arbitrary scale 0–100)			
2	30[a]	0	90[c]	0
1.5	30	90[b]	60	0
1.0	30	0	0	90[d]
0.5	30	0	0	90
0.0	30	0	0	90

[a] General low readings across the wall would indicate condensation.
[b] High readings at 1.5 m would suggest that replastering on a wall had been carried out up to 1 m but the dampness had risen to 1.5 m; replastering had not been carried out sufficiently high up the wall.
[c] Penetrating dampness from a high level is affecting the wall.
[d] Typical moisture profile suggesting rising damp to a height of between 1 and 1.5 m. But this diagnosis should be confirmed by undertaking the further analyses described below.

Timber inspection

When carrying out a survey of a building for dampness, it is often necessary to ascertain actual or potential sources of dampness, inspect any timbers that may have been affected, identify dampness defects, and assess the risk factors involved (Douglas & Singh, 1995). Much of the structural timber in any building is easily accessible, or can be exposed by lifting floorboards.

Particular difficulties arise with timber partially embedded in masonry (e.g. joist ends, rafter feet, and wall plates). Special problems occur with timber that is totally concealed by renders, plasters or wall linings, as is the case with bonding and framing timbers, and 'safe' lintels (Douglas & Singh, 1995).

The moisture meter readings may provide an indication of the need for exposure of such timbers. In some cases, and always where rot is present or suspected, opening-up of such timbers is essential. Any signs of fungal growth (i.e. mycelium or sporophores), or timber discoloration, softening, weakening or loss of strength, and any indication of insect activity, frass, or holes in timber, should be noted.

It is sometimes necessary to determine the interior condition of timber, especially in older buildings where large-size timbers such as bressummer beams are frequently encountered. Such timbers should be probed for soundness with a sharp tool, such as a knife or screwdriver. Ultrasonic sensors can be used to determine the soundness of such large members (Hollis, 1991). If the interior condition has to be assessed more accurately, an incremental borer from underneath the member and at a 45° angle into the built-in end enables samples of the core to be extracted and examined.

A fairly recent innovation in the detection of decay in wood is the use of a hand-held micro-drill supplied by companies such as Sibert Technology. This instrument allows for the non-destructive analysis of the condition of wood in service, by drilling a 1 mm diameter probe into the wood at constant pressure to depths of 200–800 mm. The rate of penetration (indicating resistance to progress) is graphically plotted. Variations in the hardness of the wood indicates changes in density, and therefore the presence of softened, decayed, or hollowed areas of timber. The extent of good and defective wood can be accurately measured from the test record (Fig. 9.2).

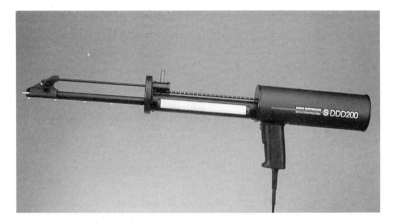

Fig. 9.2 Micro-drill for detecting wood decay (Sibert Technology).

Timber frame houses

In the design of modern timber frame houses, the use of vapour checks to prevent interstitial condensation in the insulated timber frame and the construction of clean, ventilated cavities of correct width are intended to prevent high moisture content and possibly decay developing in the timber frame itself. Uncertainties exist, in the construction of a building, as to whether these details are correctly completed (Freeman *et al.*, 1983). In addition, leaks and plumbing defects could lead to high moisture levels developing (Covington *et al.*, 1992). The problems of insulated timber frame houses have been referred to in Chapter 6.

Thus the surveyor needs to determine, as non-dest
sible, the moisture levels in the timber frame. The use o
inserted into pre-drilled holes penetrating into the sill
has been proposed, and a diagnostic service involvir
samples for laboratory examination for decay and presence of preservatives as well as moisture content is available from the BRE.

Other laboratory trials have suggested that the insertion of long nails,
up to 100 mm, which would penetrate through skirtings, plaster and the
vapour barrier and would enable readings to be taken deep in the timber
frame, would provide a method of establishing whether high readings
were present in the depth of the framework. They would not establish the
cause of this moisture, but would at least justify carrying out further
opening-up measures. Provided low readings were obtained when they
were tested with a moisture meter, these nails could be left in situ for
future inspection as required.

The most vulnerable elements of the timber frame for potential
dampness are the sole plate and sill at ground level, particularly in
kitchens and toilets (RICS Building Conservation Group, 1993). At first-
floor level, the bathroom framing is probably at the greatest risk,
because more vapour is created there than anywhere else (Covington *et
al.*, 1992, 1995).

Laboratory analyses

Salts

When any porous building material becomes wet, soluble chemicals are
normally present that either may be characteristic of the source of the
dampness or may naturally occur in the building materials themselves.
These may diffuse or spread, or be carried to other saturated materials in
the structure. Once the water vapour drops significantly, many of these
dissolved chemicals become deposited as salts, which may be characteristic of the sources or materials from which the dampness originated.

The modern methods of chemical analysis enable these salts to be
identified both qualitatively and quantitatively, and, by experience, their
presence enables detection of their origin of the moisture that deposited
them. The presence of amounts of nitrates, chlorides, sulphates and
ammonia are particularly characteristic (Table 9.5). High sulphate
contents (5–20%) would confirm sulphate attack in cementitious render,
plaster or mortar.

Table 9.5 Salt contents in water found in building materials (after Kyte, 1987).

Source	Chloride	Nitrate	Sulphate	Other
Groundwater (rising damp)	12[a]	50		
Tapwater (Buckinghamshire)	18	50		
Rainwater (Buckinghamshire)	<1	1	4	
Seawater	18 000	10		
Condensation in flues	1.2%	0.4%		0.1% ammonia
Urine contamination (human)	High	High		Ammonia sometimes
Rook leakage (tiles)			Present	Magnesium sulphate
Domestic waste			Present	Phosphates
Bricks	0.01%	0.01%	0.05%	

[a] Parts per million (ppm) unless quoted as percentage.

Plaster and render

It can sometimes be necessary to determine the composition of render or plaster that has been applied to a wall, often as part of a treatment for rising or penetrating dampness.

Chemical analysis of calcium, silica and acid-insoluble aggregate is necessary to determine whether the correct constituents of either renovating plaster or sand and cement render have been employed. Sand : cement ratios can be established, and these, alongside hygroscopic salt and sulphate analysis, help to establish the reason for any failure.

As has been described in Chapter 3 (Fig. 3.2), the correct grading of sand is an essential aspect of plaster or render formulation. The presence of too high a proportion of fine particles can lead to high porosity and premature failure.

The cementitious content of the render is first broken down chemically. The passage of the sand constituents of the sample through sieves using the BS 4551 method shows the proportion of each particle size, and is a clear way of establishing the properties and constitution of the mix used. The result of such analyses can be compared with those recommended in British Standards (Fig. 4.2).

A summary of the features that can be used during a site inspection to diagnose the sources of dampness is listed in Appendix F.

Analysis of timber decay and preservatives

Timber identification

The identification of timber can sometimes be helpful in a number of respects. Certain timbers have a heartwood that is naturally durable, and can therefore be expected to withstand the effects of excessive dampness in buildings, and some woodboring insects can affect only certain species of wood (Table 4.5). Timbers can vary in permeability to liquids and wood preservatives (Table 4.7). For all these reasons, the identification of the timber's species may be useful. With hardwoods, this can often be achieved on site with the use of a sharp razor and a hand lens. Softwoods, however, require the use of a microscope to confirm identification.

The choice of adhesive used in the manufacture of wood-based sheet materials, either plywood or particle board, influences resistance to dampness considerably. While moisture-resistant boards are available, many boards have been used that cannot withstand excessive moisture for any length of time, and they gradually deteriorate or disintegrate. The tests that are used to differentiate moisture-resistant materials are described in BS 5669. All require laboratory testing involving exposure to water and either measurement of any subsequent movement or testing for any reduction in strength.

Identification of wood decay

It is usually possible to confirm identification of timber decay on site. Certainly, dry rot is fairly easy to diagnose when it is in its advanced stage. It is not, however, so easy to distinguish between dry rot and wet rot in their initial stages of growth. Some types of wet rot can be difficult to identify precisely, and a microscope examination is helpful with certain species (Bravery, 1994). The four key factors used in distinguishing between dry rot and wet rot are: mycelium, decaying wood, strands, and sporophores (Douglas & Singh, 1995).

Frequently, however, the actual species of wet rot involved is not important, for the therapy required is usually the same regardless. The main objective is to avoid misdiagnosing wet rot instead of dry rot, or vice versa, as the treatment in each case is different. If in doubt, it may be prudent to err on the safe side by assuming that the decay has been caused by dry rot, and to treat it accordingly (see Appendix J).

Sometimes cases arise where an estimate of the longevity of an outbreak of decay is required. This is a difficult question to answer precisely, because, while the spread of growth of fungi in the laboratory under ideal conditions is known, in practice this growth may be much slower for various reasons (e.g. lower temperature conditions). It is also possible for a large outbreak to develop as a result of the amalgamation of multiple smaller outbreaks. If an exact source of dampness can be differentiated, it is safe to assume that dry rot will grow at the rate of up to 1 m per year.

Identification of the damage caused by wood-boring insects is possible in many cases by:

❑ inspection of the damage to the timber;
❑ the size and shape of the tunnels in the wood;
❑ the characteristics of the exit holes;
❑ the type of bore dust.

With the aid of a × 10 hand lens, most of the common woodboring species can usually be identified. Other household insects can usually be distinguished using that type of magnifying glass.

Timber preservation treatments

It may be useful if not important to establish whether timbers affected by or near dampness have previously had any preservative treatment. The two main traditional types of wood preservatives which have been commonly used for pretreatment of building timber are waterborne pressure systems and organic solvent double vacuum systems.

The waterborne pressure systems are usually based on copper, chromium and arsenic salts. Qualitative on-site tests are seldom required because these treatments stain the timber green. The depth of penetration is normally quite clearly defined by this colour difference as well.

Laboratory analysis can be carried out to determine the level of preservative salt retained in the treated zone. New standards and treatment specifications have been developed that lay down minimum levels of preservation in the treated zone (see BS 4072, BS 5707, BS 5268 and BS 5589).

Organic solvents may use TBTO or PCP as fungicidal components and Dieldrin, Lindane or pyrethroids using permethrin as insecticides. These preservatives do not colour the timber, and unless pigment additives are used, on-site identification of such treated timber is difficult. Borate derivatives are becoming increasingly more popular because they are more environmentally friendly.

Assessing remedial measures

Timber decay treatments

The traditional approach versus the environmental approach
Broadly speaking there are primarily two, quite opposite, approaches to dealing with timber decay as one of the main adverse effects of dampness: the traditional (major surgery) approach and the environmental (least intervention) approach. Both have their strengths and weaknesses. Some experts, however, have questioned the necessity of the traditional approach as a routine way of dealing with timber decay that may seem at times an overkill (Hutton *et al.*, 1991; Ridout, 1992).

The major surgery approach seems to be based on the premise that the only way to prevent a further outbreak is not only to cut out infected timbers but also to ensure as far as possible that any residual mycelia are either isolated or killed. This in consequence normally involves extensive removal of affected timbers and the use of substantial quantities of biocides on and in the elements concerned. The public are more aware now that many of these biocides are by nature poisonous and more or less deleterious to those exposed to them during and after the work is completed.

The second approach takes a much more modern and environmentally friendly approach to the eradication of timber decay. It seems to be based on the reasonable conclusion that timber decay in properties was not caused by a deficiency of pesticides but by building failures (Hutton *et al.*, 1991). Thus this approach considers that if the sources of moisture and infection are eliminated from the affected elements, then any fungal decay within or near it will eventually die off, provided precautionary measures such as increased ventilation are instigated soon afterwards. This approach is less intrusive and more environmentally sensitive than the traditional method. Some of the leading proponents of this approach have claimed that:

> 'By avoiding destructive exposure and radical cut back techniques, it is possible to avoid extensive damage or destruction of materials and finishes. It is also possible to avoid the use on potentially hazardous and environmentally damaging pesticides and their consequential legal and management complications.' (Hutton *et al.*, 1991)

On the grounds of safety and performance the environmental approach seems to offer an effective therapy for the eradication of timber decay. There has been no indication to the revisers' knowledge of further major outbreaks of timber decay following this approach.

The appropriate remedy to be chosen will ultimately be determined by economic and technical considerations. Generally, the environmental approach is more suitable for larger, older buildings, where conservation of the structure and fabric is a key requirement. In the majority of cases, involving smaller buildings and outbreaks, the traditional approach is likely to prevail.

Nevertheless, there may be situations where major surgery is the only feasible way of ensuring a totally successful treatment for dry rot eradication. It is appreciated, too, that guarantees will probably be harder to come by using the least-intervention approach (see below).

In-situ remedial treatment

If untreated timber with only a low level of natural durability is used, then persistent moisture content increases above 20% will inevitably lead to fungal decay. Woodworm can develop as well, and some species can attack susceptible timber even at a low moisture content.

Preventive or remedial treatment procedures are available to stop this situation developing or as part of remedial eradication measures.

Remedial treatment preservatives

The long-established procedure for remedial in-situ treatment of timber involves the spray application to accessible surfaces of structural building timber of an organic solvent-type preservative. The preservative solution is formulated to contain insecticidal and fungicidal active ingredients (Table 9.4) whose function is to eradicate as far as possible any existing infestation and to prevent any further deterioration of timber either by fungal decay or woodworm.

Over recent years, many specialist treatment companies have adopted water-based emulsified formulations, mainly for the control of insect infestations, for the obvious advantage of their lower cost, lower fire risk and lower odour compared with solvent-based products (Table 9.6). Recent research has indicated that water-based products are adequate for the control of common furniture beetle infestation in building timber. This is despite their evidently lower penetration into the timber and their lower 'initial kill' of larvae that are near the surface of the remedially treated timber (Berry, 1994).

It has to be appreciated that spray treatment, even with an organic solvent preservative, can achieve only a very modest penetration into the timber: only a few millimetres laterally on permeable timber at best. Therefore for spray treatment application to be as effective as possible as an in-situ method, the treatment must be prepared as follows:

Table 9.6 Remedial treatment preservatives: contrasting properties of organic solvent and water-based systems.

	Organic solvent	Water-based emulsions
Fire risk	Higher	Lower
Odour	Higher	Lower
Cost	Higher	Lower
Effectiveness	Higher 'initial kill' of larvae	Lower
Spray application	Better penetration into wood	Mainly on the surface
Performance and usage	Preferred for more 'hazardous' situations	Satisfactory for *Anobium punctatum* (furniture beetle)

(1) Remove insulation and sufficient floorboards as necessary to provide adequate access to joists and other timbers.
(2) Remove all bark, paint and severely damaged timber (defrassing).
(3) Remove dust, dirt and any other deposits that might interfere with the treatment.

Treatment of plywood and other glued wood composites is limited by the barrier represented by glue lines, and treatment of painted or varnished surfaces will be similarly limited.

The objective of such spray treatment is to provide a layer on the timber that will deter insects from laying eggs, and that will prevent the successful establishment of newly hatched larvae. This layer should also prevent emergence of adult beetles, though higher loadings of insecticide on the wood surface are needed in this role. Sometimes emergence through treated layers by adult insects does occur, but it is probable that successful mating and egg-laying by such creatures would be much affected.

In many situations, especially where fungal decay is involved, superficial spray treatment, even with adequate preparatory work, will need to be augmented with other in situ treatments achieving deeper penetration. One technique involves the use of plastic injectors, inserted into pre-drilled holes in the wood (Fig. 9.3). The preservative is pumped through the injector and into the wood at a depth of over 25 mm. Substantial penetration of 100 mm or more along the grain can be obtained with this technique.

The use of boiled 'mayonnaise' paste, in which the preservative active ingredients and solvents are emulsified into water incorporating a

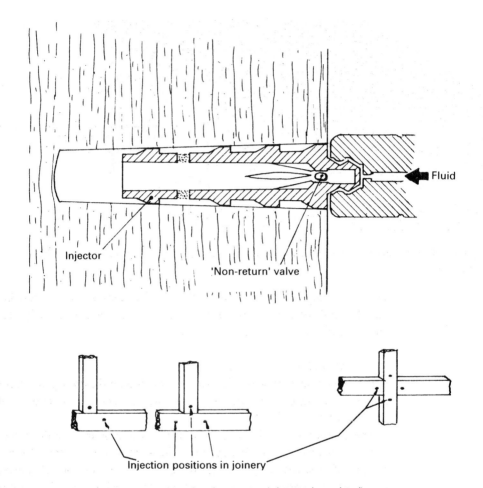

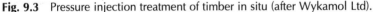

Fig. 9.3   Pressure injection treatment of timber in situ (after Wykamol Ltd).

'bodying' agent such as carboxyl methyl cellulose, has become very popular. The formulation provides a 'jellified' form of the preservative of sufficient solidity to enable it to be applied as a heavy application to accessible surfaces of timber in situ (Fig. 9.4). It enables a high loading to be achieved near vulnerable situations or where areas are inaccessible to other treatments (e.g. bond to framing timbers or exposed ends of joists in potentially damp walls). This method is more common in conservation schemes, where large timbers that can be saved require treatment.

Boron is a well-established fungicide, which has been used as a timber pre-treatment chemical for a large number of years in many parts of the world (Richardson, 1994). Like glass, it can be fused by heat into a solid, and in the form of a short rod (25 mm by 6 mm in diameter) can be inserted into holes drilled in vulnerable timbers. Boron is a very water-

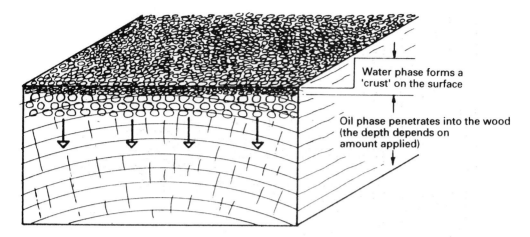

Fig. 9.4 Bodied mayonnaise-type treatment of timber in situ, showing mode of penetration into wood.

soluble chemical, and if the treated wood becomes wet it will quickly dissolve and diffuse into the water and into the wood, effectively preventing the development of any decay.

Remedial treatment of wet rot

If decay in timber is identified as wet rot, the eradication procedures involve two aspects: identification of the sources of moisture, and removal of decayed timber.

The procedures for the diagnosis of the various sources of dampness were discussed earlier in this chapter. The identification of the different species of wet rot fungus is often quite difficult on site. However, generally this is less important than the crucial decision as to whether it is wet rot or dry rot that is responsible. Some basic information on this diagnosis is to be found in Table 4.4.

Once the identification is confirmed and the sources of dampness identified, remedial measures should ensure the elimination of the associated dampness and the drying-out of the affected areas, the removal of all decayed timber and replacement, preferably with preservative pre-treated timber. Replacement timber should be physically isolated from any potential sources of dampness by the use of dpc's or other suitable impervious barriers.

Frequently the adjacent areas of sound timber close to a wet rot decay outbreak will receive a preventive spray treatment, possibly reinforced with pressure injection, timber paste or boron rod treatments wherever deeper penetration or higher preservative loadings are required.

In cases where the wet rot is localised as, say, in a timber window sill, a 'plastic' repair method (Fig. 9.5) may be preferable to partial or complete replacement of the affected section (Fig. 9.6). The former consists of cutting out all decayed wood back to a sound substrate. The exposed sound wood should be coated with a wood hardener/preservative resin. To ensure a proper key for the filler, copper nails should be inserted into the wood below the finished surface (Fig. 9.5). Once the hardener is set (usually after 24 hours), a resin/plaster paste mix is used to fill the section previously occupied by the decayed wood. After a few hours the filler should be hard enough to sand down and be primed.

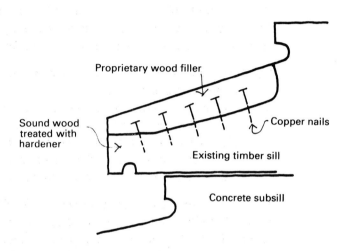

Fig. 9.5   Plastic repair method of timber affected by localised wet rot.

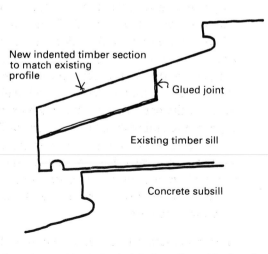

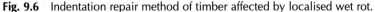

Fig. 9.6   Indentation repair method of timber affected by localised wet rot.

Remedial treatment of dry rot

Dry rot represents a particular problem in remedial treatment. There are several reasons for this:

❏ It can grow in buildings under lower moisture conditions than wet rot.
❏ It can generate additional moisture under 'static' environmental conditions.
❏ It can spread through damp masonry, decaying hidden fixing, bond and framing timbers as it grows.

Often the walls affected are substantially constructed – perhaps 300–500 mm thick – and may take a long time to dry out. BRE Digest 299 describes two stages of dry rot eradication as follows:

(1) Primary measures: locate and eliminate sources of moisture and promote rapid drying out.
(2) Secondary measures: determine the full extent of the outbreak; remove decayed wood; wall sterilisation; in-situ treatment of adjacent sound timber; use pre-treated replacement timber and support measures such as the use of damp-proof membranes to isolate timber from the sources of dampness.

There have been some recent criticisms of the widespread use of fungicidal wall sterilisations in dry rot eradication treatments (Bravery, 1991; Ridout, 1994). Wall sterilisation consists of spraying and irrigation of infected masonry with a water-soluble fungicide. The objective is to provide a chemical barrier in the wall to prevent the further spread of any deep-seated inaccessible fungus. This is in effect a 'toxic box' or *cordon sanitaire* around the outbreak. Fungicidal wall-sterilising fluids consist of aqueous solutions of fungicides such as sodium pentchlorophenate or tributyl tin oxide with quaternary ammonium compounds. Chemical barriers on the wall surface are augmented by drilling holes in the masonry and injecting the fungicidal wall solution to provide greater depth to the barrier across the treated surface, and particularly at the edges of the outbreak in order to prevent further spread.

The objections to the *cordon sanitaire* approach are compelling. They are generally held by those who favour the environmental approach rather than the major intervention approach to timber preservation referred to in Chapter 4. In summary their main objections are as follows:

❏ It introduces more moisture into the walls, which should be dried out as quickly as possible. Complete saturation would take many weeks to achieve because of the large volumes of fluid required. This would take a long time to dry out properly.

❑ The salts of the active ingredients may cause efflorescence.
❑ The chemicals may be potentially hazardous to health, although they have to be approved under current procedures.
❑ The treatment may be ineffective because of the physical difficulty of achieving a thorough and comprehensive treatment of the wall.
❑ The adequate application of these wall solutions involves stripping off plaster to expose the wall to a distance of at least 300 mm beyond the last sign of infestation of the dry rot. This is especially unacceptable if decorative or irreplaceable plaster is involved.

Despite these sound arguments, not everyone is in favour of such an approach. The contrary views, very popular within the timber preservation industry, are as follows:

❑ Wall sterilisation is essential to ensure that during the drying-out period, which may be months or even years, conditions do not still favour dry rot growth.
❑ Contractor's liability under guarantee and negligence may justify a 'safety first' policy.
❑ Costs of dealing with a recurrence of dry rot are considerable, and the risks are unacceptable to many property owners, who are expecting an instant and permanent cure.

No doubt these commercial considerations will dictate that wall sterilisation treatments will continue to be employed in many cases. However, the environmental, financial and practical drawbacks highlighted may curb the wholesale adoption of this approach.

The procedures for examining and surveying buildings for timber decay were dealt with earlier in this chapter. Outline specifications for the remedial treatment of fungal decay and woodworm are given in Appendix J.

Health and safety

As previously indicated in Chapter 4, many of the wood preservatives can be harmful. In addition, some may constitute a fire hazard while solvents are drying out.

All wood preservatives are regulated under the Control of Pesticides Regulations 1986 (via the Food and Environmental Protection Act 1985). Approval may be for general use or restricted to professional users. Standards for safe use are described briefly on product container labels and more comprehensively on manufacturers' safety leaflets.

Under the Control of Substances Hazardous to Health (COSHH) Regulations, professionals have a responsibility to carry out a risk assessment of hazardous products that will be used by an operative during treatment work. Recommendations for safe use include safety precautions to be followed by operators both for their personal protection and for that of the occupants of the buildings where treatment is carried out.

❑ Personal protection
  ■ Wear appropriate protective clothing, including: clean overalls; suitable and appropriate footwear; gauntlets to protect hands and arms; helmet to protect the head, especially when working in roof spaces; face/mouth masks; eye protection such as goggles or visors, especially when cleaning and spraying.
  ■ Ensure regular washing before meal breaks and at the end of the working day.
  ■ Use barrier cream to protect the face, arms and hands.
  ■ Use safe electrical and lighting equipment with safety cut-out switches.
  ■ Isolate or protect all electrical wiring in the building.
  ■ Extinguish all naked flames and pilot lights.
❑ Protection of the building occupants
  ■ Provide maximum ventilation of the treated areas.
  ■ Cover water tanks.
  ■ Avoid contamination of food or food storage areas.
  ■ Restrict access to the treated areas until the solvents have dispersed.
❑ Environmental protection
  ■ Use a coarse-spray jet with the appropriate pump pressure.
  ■ Arrange safe and secure storage of pesticides.
  ■ Provide safe transport of pesticides.
  ■ Provide safe disposal of empty containers.
  ■ Use correctly labelled containers.
  ■ Have fire extinguishers readily accessible.

Bats

Background
Under the Wildlife and Countryside Act 1981, certain plants and animals are designated as protected species. There have been reports that bats roosting in the roofs of buildings that have received a remedial treatment using lindane (a known hazardous chemical) as the insecticide have been injured. As a result, treatments using permethrin or other less toxic

insecticides have been recommended by the Nature Conservancy Council (NCC). If bats are found to be roosting in buildings where insecticidal or other preservation treatment of the roof timbers is necessary, the advice of the NCC should always be sought before such work is carried out.

Importance of bats

There are some 15 species of bat in the UK, and nearly 1000 worldwide. They are the only mammals that can fly, and are relatively harmless. Numbers have declined considerably in recent years, and this resulted in their being given protected status under Sections 9 and 10 of the Wildlife and Countryside Act 1981.

All British bats are insectivorous. They are active during the summer months, when one bat might consume over 3500 insects in a night, but hibernate during the winter. They roost in buildings, especially during the summer, when they may be found in cavity walls, behind weatherboarding and tile hanging, under loose flashing and behind fascia and bargeboards. Some species prefer clean dust-free buildings, and do not like cobwebs or old barns. They have several roosts, and during the summer months may use several of them during the night. Colonies may consist of as many as 500 individual bats. They like felted and insulated roofs.

Bats navigate by ultrasonic echo-location. Exposure to some remedial chemicals by either contact or inhalation may be fatal.

Recognition

When bats are actually not seen, their proximity can be inferred by the presence of mouse-like droppings, which have a distinctive point at both ends. When fresh, they are covered by a sticky glaze, and they are accompanied by portions of dismembered exoskeletons and wings of insects. Droppings are usually to be found near points of entry to a roof space.

Action

The best course is to seek the advice of the local office of the NCC. It is believed that the use of permethrin has been permitted for the treatment of infested timbers in roofs where bats may roost, although treatment is not permitted if bats are actually roosting in the roof.

Conclusions

In this chapter, the procedure for on-site surveys for dampness and the value of the instruments and the procedures that are available have been discussed. Also, some of the therapies or responses have been evaluated.

For straightforward investigations the methods mentioned are usually sufficient for a preliminary diagnosis. Naturally, recourse to laboratory analysis may be necessary to elucidate difficult or contentious situations, and it is the only appropriate method with some materials and certain situations. It is important to bear in mind that each investigation is unique, and must therefore be considered on its own merits. Checklists, equipment, and standard procedures are all helpful in assisting the surveyor to achieve a correct diagnosis. However, the surveyor must be aware of the limitations of these aids and employ them accordingly.

In summary, a number of general points can be made about the diagnosis and treatment of dampness in buildings:

❏ Before embarking on any dampness remedial work it is imperative that the actual rather than apparent source of moisture is identified and eliminated if not controlled. As has been reiterated by Howell (1995), for example, some surveyors can easily mistake condensation for rising dampness (Fig. 9.7).

❏ Moisture meters should not be relied on exclusively to diagnose dampness, particularly in masonry materials. These instruments can inadvertently give misleadingly high moisture readings. Such findings should prompt further investigations to confirm or refute the presence of dampness is present (see Appendix M).

❏ Extreme care must be used when comparing the moisture contents of different materials (Oxley & Gobert, 1994).

❏ Rising damp is not the most significant cause of moisture-related problems in buildings. In fact it is relatively rare. If present, however, it is usually persistent, while condensation is often seasonal.

❏ If you can't see it (i.e. if there are no tide marks or damp stains) it probably isn't rising damp (Rickards, 1986).

❏ Condensation in the UK is usually worse between November and March. It results in obvious defects: mould growth on walls and ceilings, damage to paintwork on the bottom rails of window sashes (Howell, 1995).

❏ Although mould is usually associated with condensation, it may also occur in some cases of penetrating damp (Garratt & Novak, 1991).

❏ The source of rainwater penetration may be well away from where the leak is appearing (Endean, 1995).

❏ Penetrating damp and leaks from services leave typical patterns of damage: efflorescence, spalling, vegetative growth and staining in the vicinity of the leak (Howell, 1995).

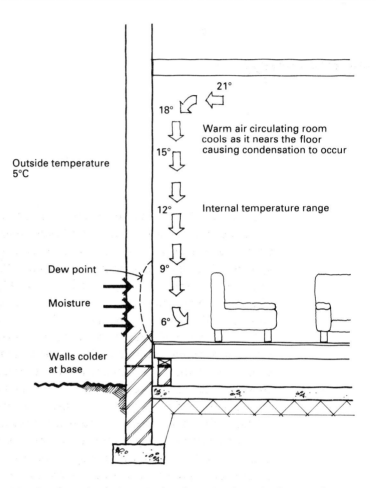

Fig. 9.7   Condensation being mistaken for rising damp (Oxley & Gobert, 1994).

❏ Moisture meters can provide some indication of the significance of dampness, but skill and experience are required in the interpretation of the readings, especially if they are used on masonry (Howell, 1995).

Postscript

The causes of dampness are many, and the remedies are nearly as numerous. It is for these reasons that investigating moisture problems in properties can be a challenging and onerous task for any building appraiser. This book has attempted to outline the nature of dampness and offer guidance on its systematic assessment. The chances of misdiagnosis can be minimised if a careful, methodical approach is taken. Not all contingencies or situations can be covered, however.

The quality of building design, the choice of materials, the activities of the occupants and the standard of construction and maintenance all contribute to today's dampness problems. Planned preventive maintenance seems to be overlooked or curbed by many property owners until problems of such severity arise that extensive repairs are essential.

Many of the dampness problems that arise could be anticipated at the design stage, particularly if a technical audit of key or vulnerable details is undertaken. This can be done by a building surveyor or some other construction professional with experience in identifying potential failures. However, some defects are the result of lack of knowledge or care by site supervisory and operating personnel. More specialist guidance and training are necessary for these workers, and greater emphasis is required on many construction courses at both technician and graduate levels on the repair and rehabilitation of buildings. After all, this sector now accounts for nearly half of all construction output in the UK. The new National Vocational courses in construction and undergraduate programmes in building surveying have gone some way to fulfil this requirement. Most of these programmes contain modules dealing with the fundamental aspects of maintenance management and technology. On the research side, British technical universities, professional bodies, and agencies such as the BRE are making great progress in increasing our understanding of building problems and in helping to improve existing and new methods and techniques. Slowly but surely we are learning from past mistakes, but in our human frailty, defects will probably never be completely eradicated. Our main aim must be to minimise their effects even if it is recognised that it may be impossible to avoid them.

Appendix A
Portland Cement

Basic manufacturing process

The basic raw materials for ordinary Portland cement are clay or shale and calcium carbonate in the form of chalk or limestone. Most clays are complex materials containing silica (SiO_2), alumina (Al_2O_3) and iron oxide (Fe_2O_3) (Table A1), but these oxides are not separately identifiable. In the manufacture of cement, a typical raw material would consist of calcium carbonate, to provide lime (CaO), and clay. These are finely ground and heated to about 1350 °C in a rotary kiln.

Table A1 Main constituents of cement (after Neville & Brooks, 1987).

Constituent	Formula	Proportion (%)
Calcium oxide	CaO	60–67
Silica	SiO_2	17–25
Alumina	Al_2O_3	3–8
Iron oxide	Fe_2O_3	0.5–6.0
Magnesium oxide	MgO	0.1–4.0
Alkalis		0.2–1.3
Sulphur trioxide	SO_3	1–3

Using a shorthand adopted by cement chemists (C = CaO, S = SiO_2, A = Al_2O_3 and F = Fe_2O_3), the processes occurring in manufacture are shown in Table A2. Dehydration of this mixture occurs at 400 °C (Table A2 (a)). At 900 °C calcium carbonate decomposes to calcium oxide and, as the temperature increases still further, dicalcium silicate (C_2S), tricalcium aluminate (C_3A) and tetracalcium aluminoferrite (C_4AF) are formed (Table A2 (c)). These fuse and react with dicalcium silicate and calcium carbonate to form tricalcium silicate (C_3S). Tricalcium aluminate (C_3A) and tetracalcium alumino ferrite (C_4AF) form a matrix around the tricalcium silicate (C_3S) and dicalcium silicate (C_2S) (Table A2 (e)), and a clinker is formed on cooling; in none of these compounds are the various oxides separately identifiable, but the compounds themselves are well-characterised chemical entities; 5% gypsum ($CaSO_4$) is added, and it acts as a retarder. The whole mixture is then ground to a fine powder known as Portland cement. The proportions and role of the different components are shown in Table A3.

Table A2 Stages of cement manufacture.

Temperature	Effect	Stage
400 °C	Loss of water	(a)
900 °C and above	$CaCO_3 \rightarrow CaO$	(b)
Dicalcium silicate	C_2S	
Tricalcium aluminate	C_3A Form and fuse	(c)
Tetracalcium aluminoferrite	C_4AF	
Dicalcium silicate	C_2S reacts with $Ca \rightarrow C_3S$	(d)
Di/Tri calcium silicate	$C_2S + C_3S$ in a matrix of $C_3A +$	(e)
	C_4AF	

Table A3 Proportions, constituents and functions of cement components.

Formula	Proportion (%)	Role
C_3A Tricalcium aluminate $3CaO.Al_2O_3$	7–10	Initial setting Reacts rapidly with water (days); retarded by $CaSO_4$
C_4AF Tetracalcium alumino-ferrite $4CaO. Al_2O_3. Fe_2O_3$	7–10	Sets quickly; retarded by $CaSO_4$ Gives cement its grey colour
C_3S Tricalcium silicate $3CaO. Si_2$	30–60	Hydrates quickly (weeks) Rapid strength
C_2S Dicalcium silicate $2CaO. Si_2$	15–35	Ultimate strength Slow hydration (months)
$CaSO_4$ (Gypsum)	5	Retarder for tricalcium aluminate

Hydration of cement

When a cement powder is mixed with water, it undergoes a process called hydration, which leads to setting and hardening. Depending on the aggregate, the familiar range of cement-based products are formed: mortar, render, concrete and concrete blocks and bricks, for example.

Immediately on mixing with water, the cement grains become dispersed in the mixing water. Their distribution will be determined by the amount of water added (water : cement ratio). The grains remain in suspension until setting. The tricalcium aluminate (C_3A) reacts very rapidly with water:

$$3CaO. Al_2O_3 + 6H_2O \rightarrow 3CaO. Al_2O_3. 6H_2O$$
(Tricalcium aluminate)
$\quad\quad$ (C_3A)

To avoid too rapid setting as a result of this reaction, which would stiffen the cement immediately water was added, the gypsum powder reacts to form ettringite:

$$3CaO. Al_2O_3 + CaSO_4 + 32H_2O \rightarrow 3CaO.Al_2O_3. CaSO_4. 32H_2O$$

This is insoluble and settles out. The gypsum is all used up in about 24 hours, and so normal setting can proceed. In fact, the gypsum also reacts with tetracalcium aluminoferrite (C_4AF) to form calcium sulphoferrite and sulphoaluminate. The hydration reaction for C_4AF is uncertain, but it is believed that lime is involved:

$$4CaO. Al_2O_3. Fe_2O3 + 2Ca(OH)_2 + 10H_2O \rightarrow 3CaO. Al_2O_3. 6H_2O$$
$$+ CaO. FeO_3. 6H_2O$$
(Tetracalcium aluminoferrite)$\quad\quad$(lime)
$\quad\quad$ (C_4AF)

However, tetracalcium alumino ferrite (C_4AF) is present as every small part of the total cement (10%) and provides a contribution to the crystalline hydration products.

The two calcium silicates produce the bulk of the hydrated material:

$$2CaO. SiO_2 + 3CaO. SiO_2 + nH_2O \rightarrow xCaO. ySiO_2. zH_2O$$
$$+ Ca (OH)_2$$
$\quad\quad\quad\quad\quad\quad\quad\quad$(calcium silicate hydrate)$\quad$(lime)
$\quad\quad\quad$ (n, x, y, z variable)

and they produce calcium silicate hydrate or cement gel in the form of particles of colloidal dimensions, whose composition is ill defined and depends on the conditions of formation. The lime contributes to the crystalline products.

At this stage the hydration on the cement grain surfaces is occurring, and electron or stereoscan microscope observations have been interpreted as indicating the development of rods of ettringite and needle-like forms of calcium silicate hydrate in the spaces between the cement grains. After 24 hours, setting has occurred but no real strength has yet developed. A continuous gel establishing a solid skeletal structure is reinforced by crystalline products like ettringite and tricalcium aluminate. Tetracalcium alumino ferrite (C_4AF) hydrates and large crystals of calcium hydroxide form.

After several days, strength has developed, although not all the

(a)

(b)

(c)

(d)

Fig. A1   Diagrammatic illustration of the cement hydration process.
(a) Unhydrated cement particles; (b) After 24 hours, rods of ettringite and calcium silicate form and begin to hydrate. No real strength has yet developed;
(c) Establishment of a continuous gel with a solid skeletal structure reinforced by crystalline products commences; (d) After several days, strength has developed, but not all hydration reactions are completed. Some pores are unfilled, and as the ageing process continues, pore volume declines, and larger pores are reduced in number. (Based on Pavers (1966), Proceedings of the 4th International Symposium on the Chemistry of Cement, by permission of the National Bureau of Standards, Washington DC.)

hydration reactions have been completed. More hydration links have developed, but some pores remain unfilled. As ageing proceeds, the total pore volume declines and the larger pores are reduced in number (Fig. A1).

Appendix B
The Manufacture of Lime

(1) Limestone is heated in a kiln to convert it to quicklime:

$$Ca\ CO_3 \rightarrow Ca\ O + CO_2$$
900–1200 °C

(2) The slaking process involves the addition of water, which can sometimes cause a violent reaction. This produces lime putty:

$$Ca\ O + H_2O \rightarrow Ca\ (OH)_2$$
Calcium hydroxide

(3) The setting of lime involves reaction with atmospheric carbon dioxide. If cracks develop in service, the freshly exposed surfaces will cure by further reaction of the freshly exposed calcium hydroxide with carbon dioxide:

$$Ca\ (OH_2) + CO_2 \rightarrow Ca\ CO_3 + H_2O$$

Gypsum Plaster Cement

Natural gypsum is hydrated calcium sulphate ($CaSO_4 . 2H_2O$), which, when heated at relatively low temperatures ($150\,^{\circ}C$), dries to form the hemihydrate $CaSO_4\, \frac{1}{2}H_2O$. This is plaster of Paris, and is described as Class A plaster in BS 1191:

$$CaSO_4 . 2H_2O \rightarrow CaSO_4\, \tfrac{1}{2}H_2O3CaO. + 1\tfrac{1}{2}H_2O$$
$$\text{1300\,}^{\circ}C \text{ in a kiln} \quad \text{hemihydrate}$$
$$\text{(plaster of Paris powder)}$$

It is used in some small repairs as it sets quickly. For more general work, the setting is retarded by the inclusion of proprietary admixtures such as keratin or calcium tartrate, which control the crystallisation reaction, the evolution of heat and the expansion of the setting reaction:

$$CaSO_4 + 1\tfrac{1}{2}H_2O \rightarrow CaSO_4 . 2H_2O$$
$$\text{(set plaster)}$$

For undercoat plaster, one part of the retarded hemihydrate is mixed $1:3$ with sand for use on brick walls, or as $1:5$ when applied to concrete. Thistle browning is a common example of this type of plaster.

If calcium sulphate hemihydrate is further heated at higher temperatures, the anhydrous salt $CaSO_4$ is produced:

$$CaSO_4 . \tfrac{1}{2}H_2O \rightarrow CaSO_4 . + \tfrac{1}{2}H_2O$$
$$\text{160--170\,}^{\circ}C \quad \text{(anhydrous parian plaster)}$$

This parian plaster is a slower-setting form, and an accelerator such as alum is added to produce a slow but continuous set. Sirapite (class C) and Keene's plaster (class D) are of this type, but these are not readily available now commercially. They are normally mixed with lime and sand to form undercoats or used neat as a finish. They form a hard impervious coat, which may attract condensation. (Sirapite is a $60:40$ anhydrous:hemihydrate plaster mixture. In the 1960s it was often used as the material for lightweight 'concrete' wall panels of some industrial buildings because of its excellent insulation properties.)

The setting of all these plasters is accompanied by expansion but, when complete, little further movement occurs. Therefore coats cannot be added before the previous coat has dried provided good intercoat adhesion can be obtained. In this respect, gypsum plasters are superior to cement-based systems. However, the former are not suitable for plastering or replastering where damp conditions have occurred or may reoccur.

Appendix D
Test Paper for Chlorides

Procedure

The test papers should be freshly prepared from filter papers by dipping or spraying them with a 0.2% solution of potassium chromate and then drying.

The dry paper is then dipped or sprayed with a 0.2% silver nitrate solution. This precipitates silver chromate on the paper, and it turns a red colour. Excess silver nitrate is washed off the paper with distilled water, and the paper is then dried.

Prepared test strips should be stored in a lightproof box.

If a dampened test paper is placed in contact with a sample containing chloride, it is decolorised.

Result

If the test paper turns a red colour when wetted with the solution, chlorides are present.

Appendix E
Hygroscopic and Capillary Moisture Analysis

Procedure

The procedure for the analysis of walls for suspected rising dampness and the determination of hygroscopic and capillary moisture is as follows.

Samples to be examined should be taken from the wall at floor or ground level to at least 300 mm above the height of the indicated dampness. The samples should be taken by drilling, preferably in the mortar courses at about 300 mm centres, to the centre of the wall. It is preferable if the inner samples are separated from samples derived from the plaster, as it is frequently desirable to analyse both the masonry and the plaster.

The samples are obtained by drilling at a low speed with a masonry bit of about 9–15 mm in diameter. High-speed drilling may cause some premature drying of the drilled-out sample. The samples are collected in airtight containers. Details of the position of the sample must be recorded.

In a laboratory, about 2 g of the sample are accurately weighed (W_w) and then are exposed in a container that has a 75% relative humidity for at least 12 hours. After reweighing (W_{75RH}), the samples are placed in an oven at 100 °C until dry, and then are reweighed (W_4).

Hygroscopic moisture content (HMC) at 75% RH is calculated as follows:

$$\text{HMC} = \frac{W_{75RH} - W_4}{W_{75RH}} \times \frac{100}{1}$$

Total moisture content (TMC) is calculated as follows:

$$\text{TMC} = \frac{W_w - W_4}{W_w} \times \frac{100}{1}$$

Capillary moisture content (CMC) is calculated by subtracting HMC from TMC:

$$\text{CMC} = \text{TMC} - \text{HMC}$$

Result

If CMC > HMC, then rising damp is present. If HMC > CMC, then the source of dampness is likely to be hygroscopic salts caused by condensation.

Appendix F
Dampness Investigation Checklist

PROPERTY: **JOB No.:**

USE Category:

SURVEYOR:

DATE OF INSPECTION: **WEATHER:**

(1) WALLS

FORM: SOLID/CAVITY/CLAD-FRAME
THICKNESS:
MATERIAL: STONE/BRICK/BLOCK
POINTING/JOINTS:
DPC: YES/NO   POSITION:
OUTSIDE FINISH: FAIR FACE/RENDER/CLAD
INSIDE FINISH: FAIR FACE/PLASTER/L&P/P-B
AIR MOISTURE: AIR TEMP:
SURFACE TEMP: DEW POINT TEMP:
ROOM HEATED: YES/NO   TYPE:
OPEN HEARTH FIRE: YES/NO   BLOCKED UP   YES/NO
ROOM VENTILATED: YES/NO
FUNGAL GROWTH: YES/NO   TIDE MARKS: YES/NO
SALTS DETECTED: YES/NO
OUTSIDE G/L POSITION ABOVE GFL:
OUTSIDE G/L FORMATION:
INTERNAL FLOOR: SOLID/SUSPENDED
SERVICES NEARBY: YES/NO
DAMP NEAR OPENING/JUNCTION/ABUTMENT: YES/NO
POSITION OF DAMPNESS:

(2) FLOOR

CONSTRUCTION: SOLID/SUSPENDED
SUB-FLOOR VENTILATION: GOOD/FAIR/POOR
FLOOR MOISTURE CONTENT:

(3)   CEILING

FLOOR CONSTRUCTION: SOLID/SUSPENDED
SUB-FLOOR VENTILATION: GOOD/FAIR/POOR
FLOOR MOISTURE CONTENT:

(4)   WALL SURFACE MOISTURE CONTENT (ROOM):

Level	Locus 1	Locus 2	Locus 3	Locus 4	Locus 5
Skirting					
200 mm					
300 mm					
400 mm					
500 mm					
600 mm					
700 mm					
800 mm					
900 mm					
1000 mm					
1100 mm					
1200 mm					

Appendix G
Dampness Diagnosis Checklist

Evidence			Type of dampness		
Form	Present yes/no	Condition	Rising damp	Penetrating damp	Other
Symptoms (1) Mould/mildew/fungi? (2) Algae/lichen/moss? (3) Tide marks/distinct damp patches? (4) Salt stains/patches? (5) Nothing obvious?					
Construction (6) Solid/cavity masonry? (7) Defective supply pipes? (8) Faulty rainwater goods? (9) High outside ground level? (10) Defective flashings/dpc's/ pointing?					
Moisture readings (11) WME of floorboards (average) (12) WME of skirtings (average) (13) Sharp changes at the interface of wet and dry? (14) WME of wall surfaces under 40%? (15) WME of wall surfaces between 40 and 70%? (16) WME of wall surfaces over 70%? (17) RH above 70%?					
Totals					

(Scoring: 2 = Likely   1 = Possible   0 = Unlikely)

Appendix H
Comparison of Dampness Symptoms

Evidence	Condensation	Rising damp	Penetrating damp
WME readings			
Moisture readings at margin (especially the upper margins of the damp areas)	Gradual change from wet to dry	Sharp change from wet to dry	Usually a sharp change from wet to dry
Moisture readings in skirting and floor in direct contact with wall	Low readings	High readings	High readings in lower part of wall affected
Moisture readings 0–40%	Possible if near 40%	No	Unlikely
Moisture readings 40–70%	Possibly	Remotely possible	Possibly
Moisture readings 70–100%	Remotely possible	Possible	Remotely possible
Moisture readings taken at various depths in the wall using deep wall probes	High at surface, lower at depth	High all through	Generally high all through. Higher towards the source and often lower towards wall surface
Symptoms			
Are there any mouldy patches?	Yes. They may be relatively dry at the time of survey	Very rarely	Sometimes
Is mould growth especially noticeable behind pictures and furniture or in corners or enclosed spaces?	Yes	No	No
Elf cup fungus	No	Unlikely	Possibly
Tide mark on wall (salts present)	No	Yes	Possibly

Evidence	Condensation	Rising damp	Penetrating damp
Symptoms (contd)			
Soil salts, including nitrate, in wallpaper or in a scraping from the wall surface	Absent	Yes	Possibly
No visible signs of dampness	Possibly	Unlikely	Possibly
Patch of saturated plaster	Unlikely	Unlikely	Probably – or leaking services or hygroscopicity
Construction defects			
Faulty rainwater goods	No	No	Yes
External ground levels at or above dpc	No	No	Yes
Cracked plinth not keyed to wall at top	No	No	Yes
Unventilated/unused flue	Yes (+ sulphates)	No	Possibly
Faulty flashings	No	No	Yes
Perished/cracked external paintwork	No	No	Yes
Spalled brickwork/rendering	No	Possibly	Yes
Narrow band of high readings above skirtings in solid walls	No	Possibly. Dpc bridged by render	Yes
Narrow band of high readings above skirtings in cavity walls	No	Possibly. Dpc bridged in cavity	Possibly
User activities			
Building vacant all day without heating or ventilation	Yes	No	Possibly
Paraffin or portable gas heaters used	Yes	No	No
Doors of bathroom and kitchen kept open, vents blocked up	Yes	No	No
Inadequate maintenance	Possibly	Possibly	Yes

Appendix J
Timber Treatment Specifications

Dry rot

(1) Establish and eliminate the cause/s of the dampness.

(2) Cut out all decayed wood to a distance of up to 1 m beyond the last sign of decay. Bag, cart away and burn all removed material.

(3) Cut away plaster and rendering, remove skirtings and skirting grounds, wall linings, architraves, bond and framing timbers on and in infected masonry to a distance of up to 600 mm away beyond the last sign of infection.

(4) Remove mycelium from exposed masonry and clean down walls with a wire brush.

(5) Sterilisation. *Either:*

    (a) Carefully blowtorch to exposed masonry surfaces. Surface-spray exposed masonry, oversite, etc. with a boron-type fungicidal wall solution. *Or*

    (b) Surface-spray exposed masonry, oversite, etc. with a fungicidal wall solution incorporating either sodium penta-chlorphenate, sodium orthophenylphenate or tributyl tin oxide with quaternary ammonium compounds; *and*

    (c) Irrigate masonry by application of the fungicidal wall solution via holes drilled into the masonry, usually in the perpendicular joints, to reinforce surface spraying to:

        (i) establish a *cordon sanitaire* around the outbreak;

        (ii) separate the base of an infected wall lacking a dpc;

        (iii) provide additional protection in other special circumstances.

(6) All replacement timber is to be treated in accordance with BS 5268 Part 5. All ends and surfaces of timber that are cut or worked after treatment must be retreated by two liberal brush or spray coats, preferably using an organic solvent preservative, before fixing. Isolate all timber from damp oversite, or damp masonry, and any other possible sources of dampness by the use of dpm's.

(7)   All existing timber in the affected area should be spray treated on all accessible surfaces with an organic solvent preservative at a rate of 1 litre per m^2 of flooring area and 2 litres per m^2 of roofing area. Where timber is at special risk (i.e. in contact with damp masonry), it should be further treated by the application of bodied mayonnaise-type emulsion (timber paste) at the rate of 0.5–1.0 kg/ 5 m^2 wood surface area. Alternatively, local pressure injection of an organic solvent preservative using plastic, non-return injectors or the insertion of boron rods would be satisfactory.

(8)   As an additional protection, the masonry sterilisation procedure can be reinforced by use of zinc oxychloride additives to plaster or as a paint on the finished surface.

Wet rot treatment specification

(1)   Establish the cause/s of dampness and effect a permanent cure.

(2)   Cut out all decayed wood to a distance of 300 mm approximately beyond the last sign of decay.

(3)   Clean the timber that will remain in situ, preferably with an industrial vacuum cleaner.

(4)   Clean up the site. Remove decayed timber from the area and burn.

(5)   Isolate all timber from damp oversite, or damp masonry, and any other possible sources of dampness by the use of dpm's.

(6)   All replacement timber is to be treated in accordance with BS 5268 Part 5. All ends and surfaces of timber that are cut or worked after treatment must be retreated by two liberal brush or spray coats, preferably using an organic solvent preservative, before fixing.

(7)   All existing timber in the affected area should be spray treated on all accessible surfaces with an organic solvent preservative at a rate of 1 litre per m^2 of flooring area and 2 litres per m^2 of roofing area. Where timber is at special risk (i.e. in contact with damp masonry), it should be further treated by the application of bodied mayonnaise-type emulsion (timber paste) at the rate of 0.5–1.0 kg/ 5 m^2 wood surface area. Alternatively, local pressure injection of an organic solvent preservative using plastic, non-return injectors or the insertion of boron rods would be satisfactory.

Treatment specification for woodworm

(1)   Lift all floorboards to provide adequate access to subfloor joists and plates. Boards next to walls should always be lifted to give access to joist ends.

(2)   Remove any insulation or other material that would restrict access to surfaces of timber to be treated.

(3)   Remove any adhering bark and severely affected timbers, and cut away any severely damaged areas. Any replacements should use preservative-treated timber. Clean down all accessible surfaces of floorboards, joists, rafters and sarking boards by vacuum cleaning or by brushing. Remove dust and debris from the treatment area.

(4)   Treat all accessible timbers by applying, by spray, insecticidal fluid at a rate of 1 litre per m^2 of flooring area. Treat both surfaces of floorboards, including those that have been lifted prior to relaying.

(5)   Staircases: treat all accessible surfaces by spray. Inject preservative into any old flight holes. If the undersides of the staircase timbers are inaccessible, holes should be drilled in the risers and a spray lance inserted to treat the soffit surfaces.

Appendix K
Typical Replastering Specification

Workmanship

(1) The replastering work shall comply with the following codes of practice: BS 5492: 1990 *Code of practice for internal plastering*; BS 8000 *Workmanship on building sites* Part 10: 1989 *Code of practice for plastering and rendering*; and BS 6524: 1985 *Code of practice for installation of damp-proof courses.*

Preparation

(2) Adequate provision shall be made to protect all floor finishes and other parts of the property which may be affected by the replastering work.

(3) All existing plaster shall be removed to 1.0 m above the line of the dpc or 300 mm above the highest point affected by salts.

(4) Clean down the wall exposed with a wire brush to remove laitance and debris.

(5) Rake out mortar joints to a depth of at least 13 mm to provide a good key (or apply a 3 mm thick splatterdash coat containing an SBR bonding agent to the entire substrate to be plastered).

(6) Delay replastering as long as possible to allow the wall to dry out If necessary, instigate rapid drying-out measures.

Application

For brickwork built in cement sand mortar

(7) Apply the first coat of cement plaster as a 1:3–4 cement:sand mix. Sand shall be BS 882 type M. The mix shall incorporate a

waterproof additive and the application should be to a thickness of 12 mm. The minimum of water shall be added to the mix. The surface shall be scratched to provide a key for the second coat.

For masonry built using a lime-based mortar

(8) Do *not* use a cement-based render. Rather, apply a first coat using a 'renovating' plaster. Sand shall be BS 882 type M. The mix shall incorporate a waterproof additive and the application should be to a thickness of 12 mm. The minimum of water shall be added to the mix. The surface shall be scratched to provide a key for the second coat.

(9) The second coat should be similar to the first, to a thickness of 8 mm, and should be applied to the undercoat before the first has finally set, to ensure satisfactory adhesion. The surface shall be combed to provide a key for the final coat.

(10) Brush off any salting appearing on the second coat finish.

(11) Apply a thin coat (3 mm) of a non-sulphate type of plaster. This must be porous and not overtrowelled.

(12) Redecoration should be in porous paints (e.g. emulsion) only.

(13) A 25 mm wide gap should be left between the plaster and the floor at the base of the wall.

Appendix L
Construction Moisture and Dampness Testing

(Extract from BS 8203: 1987 *Code of practice for installation of sheet and tile flooring*, with permission)

Eliminating construction moisture

Before a floor is laid it is necessary not only to ensure that the floor is constructed to prevent moisture reaching it from the ground but also to ensure that sufficient of the water used in construction is eliminated. Usually the flooring is fixed directly to the concrete base slab or onto a screed laid above this. In either case the amount of water used is more than that required for hydration of any cement and because extra is normally required to give adequate workability to the mix. It is essential that the excess water be allowed to evaporate and the time for this to happen should be taken into account at the planning stage. Estimated drying times are necessarily only very approximate as drying is influenced by ambient conditions, concrete quality, surface finish and thickness. Of these, thickness is the most important. For the first 50 mm one day per mm should be allowed followed by an increasing time for each millimetre above this thickness. It is thus reasonable to expect a screed 50 mm thick laid on a membrane to be sufficiently dry in two months. However, concrete 150 mm thick may require as much as a year to dry from one face only. Where screeds are laid directly onto a concrete base, account should be taken of extra drying time required for the base.

Whenever time schedules do not permit the extended drying time for thick concrete bases, an unbonded fine concrete or cement sand screed thick enough to provide sufficient rigidity and to minimise the likely curling with the damp-proof membrane placed between the slab and the screed should be used.

The flooring should not be laid until a hygrometer test by the method described below gives a reading which indicates a relative humidity of 75% or less.

Electrical methods of conductivity measurement (see below) provide a means of monitoring the progress of the drying-out operation. However, the final assessment of dryness should be made using the hygrometer.

Chemical methods of assessing dryness are not recommended.

Hygrometer test

Basis of test

The recommended method for dampness testing in 11.3 of BS 8203 is by use of the hygrometer. Concrete under normal conditions will never be completely dry. Those responsible for laying flooring require to know when the moisture level of the concrete has been reduced to a value where flooring can safely be laid. Water in the coarse pores of concrete is relatively mobile and can lead to damage to flooring whereas water in fine pores is relatively immobile and harmless. When concrete is allowed to dry, the coarse pores become empty first because water in coarse pores exerts a higher vapour pressure than in fine pores. Because the size of the pores controls the vapour pressure that arises in them, it also controls the vapour pressure of a small volume of air entrapped between the concrete surface and the hygrometer. The vapour pressure determines the relative humidity of that entrapped air so a hygrometer reading indicates the extent to which harmful moisture is still present. Experimental evidence has shown that when moisture has evaporated from the coarse pores the relative humidity falls to 80%. If some allowance is made for errors in determining the relative humidity, it is considered reasonable to recommend that the concrete be considered dry when the relative humidity falls to 75% or less.

For these reasons the hygrometer method for dampness measurement is recommended.

Apparatus

Any instrument which is capable of measuring the relative humidity of a pocket of air trapped between the base and in contact with it can be used. The pocket of air should be isolated from the atmosphere by thermal insulation and a vapour barrier.

A suitable form of apparatus is shown in Fig. L1. This is made from

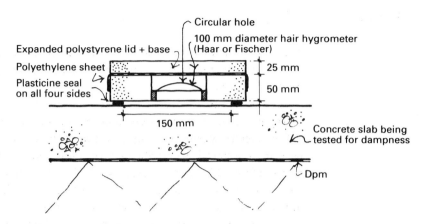

Fig. L1   Hygrometer on concrete floor slab.

readily available materials. It can be made from expanded polystyrene sheet through which a hole is cut in the centre. The sheet should be sufficiently thick to house a hygrometer which can be fixed into the foam or placed on the floor. A sheet of polyethylene (500 gauge is suitable) is stuck to the top of the foam to form a vapour barrier and to close the hole. A loose lid consisting of a further layer of expanded polystyrene completes the apparatus.

The lid can be lifted and a reading taken without disturbing the pocket of air.

The apparatus can be sealed to the floor; and a material such as Plasticine is convenient for this purpose.

Either paper or hair hygrometers may be used. The latter are, in general, more accurate and stable so need calibrating less often.

NB: Electronic instruments which measure relative humidity are also available but insufficient experience has been gained with them at present to assess their use.

Verification of hygrometer

The accuracy of a hygrometer reading can be subject to drift with time and the instrument therefore needs to be recalibrated frequently. As the important relative humidity is 75%, the hygrometer may be checked by sealing it above a dish containing a saturated solution of sodium chloride for at least half an hour. The atmosphere above the solution will be 75% over the complete temperature range in which the instrument is likely to be used.

Procedure

Turn off any artificial aids used for accelerating drying at least four days before final readings are attempted. Accelerated drying should not be used for screeds.

Seal the instrument firmly to the floor and allow sufficient time for the entrapped air to reach moisture equilibrium with the screed or base.

For screeds, where the damp-proof membrane is placed between the base and the screed as described above, allow a period of not less than 4 hours before taking the first reading. Equilibrium can be assumed either when two consecutive readings taken at hourly intervals show no change or if the instrument is left in position overnight.

For very thick constructions, i.e. where the damp-proof membrane is placed below the base slab, allow a period of at least 72 hours to elapse before taking the first reading. Equilibrium can be assumed when two consecutive readings taken at four-hourly intervals show no change. Construction thicknesses greater than 200 mm can take considerably longer than 72 hours before moisture equilibrium is established. To prevent edge effects with these very thick constructions, the area for $0.5\,m^2$ surrounding the instrument should be covered with an impervious sheet material during the test.

To minimise the time for the instrument to be in a position on the floor the following technique can be applied. Cover the positions to be measured with impervious mats (e.g. polyethylene sheet, rubber mats) not less than 1 m × 1 m, taped to the floor at their edges. Leave in position for at least three days in the case of screeds and seven days in the case of thick constructions. After removing the mat, immediately seal the instrument to the centre of the covered area. Experience has shown that moisture equilibrium is usually attained within 2–4 hours of placing the instrument.

Conductivity test

A quick way of obtaining an approximation of the moisture content of sand and cement screeds or concrete is to measure the electrical conductivity. A suitable instrument for this purpose is one which is particularly designed to use the gel bridge method for the measurement of conductivity between two electrodes which have been inserted into two pre-drilled holes 25 mm deep in the concrete sub-floor.

The presence of chlorides, whether present as deliberate additions or otherwise, and other screed additives can give rise to erroneous readings.

The error will depend on the quantity present but, in general, the water content indicated by the tests will be the maximum water content.

Care should be taken in carrying out the test to avoid contact with the probes and any metal embedded or incorporated in the screed.

Although a high reading indicates wet concrete, in some instances, e.g. where a damp-proof membrane is placed below the base slab, or a power-trowelled slab or thick screed has been specified, a low reading does not necessarily indicate satisfactory dryness; below the surface the concrete may be still wet.

In these cases extra checks should be made by taking readings at a number of places, which are then covered by an impervious material, e.g. polyethylene or glass. After 24 hours the covers should be removed, new holes bored and further readings taken at each place. Where the second reading significantly exceeds the first it should be taken as an indication that the concrete is not sufficiently dry.

Appendix M
Guidance for Surveyors in Measuring Moisture

The following is a list of important points on assessing moisture in masonry made by Jeff Howell of South Bank University at a major UK property research conference (Howell, 1995). Surveyors would do well to heed his advice.

(1) Be aware that the walls of older properties must be expected to have a certain moisture content. A totally 'dry' internal decorative surface can be achieved only by use of an impermeable plastering system. This will not result in a 'dry' wall, but will only trap water behind it. This may upset the moisture equilibrium of the structure.

(2) Beware of recommending 'further investigation by specialists'. Most damp-proofing companies are not interested in specifying rectification of condensation, rainwater or plumbing or drainage problems. They specify only the work that they themselves carry out, which is chemical damp-proofing. Most 'specialist' surveyors rely solely upon electric conductance meters. They are often paid on a commission-only basis.

(3) If you use an electrical conductance or capacitance meter to carry out surveys, make sure you are aware of its limitations. Take and record readings all around the building, not just at low level. Take readings up the full height of the wall. Uniform high readings or high readings above about 1 m are unlikely to be due to capillary moisture.

(4) If you suspect that condensation may be a problem, take RH and surface temperature readings. Surface temperature readings within 5°C of dew-point should be especially noted.

(5) Be suspicious of high electrical conductance readings in 'wet' rooms or unventilated/unheated rooms. Be suspicious of high readings in the vicinity of windows, doors, rainwater goods, water service pipes and drainage pipes.

(6)   Always be aware of the potential for damage to the building caused by unnecessary work, especially replastering. There are a number of client actions pending regarding work carried out without evidence of correct diagnosis.

Glossary

Absolute humidity Measure of the actual quantity of water vapour present in the air at any given time. It is a variable quantity measured in kg/m^3, and can best be found by experiment using a glass jar and a 'U' tube containing calcium chloride.

Absorption The penetration of a substance such as water into the body of another.

Additive Strictly speaking, an additive is a material that is added as part of the cement or other binder during its manufacture. Examples of such additives would be cement process grinding aids such as polyethylene glycol. However, the term is also used to describe ingredients such as plasticisers, waterproofers and other special agents that are pre-mixed with the cement or plaster prior to its being bagged.

Admixture A special material added in small quantities to cement or concrete mixes, which alters the properties or strength of the product. Examples of admixtures are air-entraining agents, or waterproofing compounds that are added during the mixing process on site. This is the essential difference between admixtures (added on site) and additives (added during manufacture).

Adsorption The taking up of one substance at the surface of another.

Air dry The condition of a material in an ordinary indoor, inhabited environment with a relative humidity not exceeding about 70%.

Air drying (natural seasoning) Converted green timber can be dried to a moisture content of about 16–18% by stacking the boards in such a manner that a free flow of air is possible over all surfaces, and the stack is protected from the rain. It can be a time-consuming process, and with some timbers the air drying rate may be as slow as 25 mm (1 in) thickness of timber drying per year. Many damp building materials may dry out as a result of being exposed to the air in a building. If they are protected from dampness, they will achieve a moisture content in balance with the prevailing ambient humidity conditions.

Angiosperms   A major group of plants with closed seeds. This group includes all hardwoods (i.e. broad-leafed trees). Compare with *gymnosperms*.

Antigen   Any substance capable of stimulating an immune response; usually a protein or large carbohydrate that is foreign to the body.

Arenaceous   Sedimentary rocks in which the principal constituents are sand grains, including the various sorts of sands and sandstones.

Beam-filling   A crude masonry infill about 600 mm high, found in traditional pitched roofs of Georgian and Victorian buildings to provide a degree of fire-stopping and draughtproofing at the eaves.

Bore dust (frass)   The dust deposited by certain wood-boring insect larvae in their tunnels. It usually consists of chewed wood fragments and faecal pellets. These pellets comprise wood residues that have passed through the gut of the larvae. They have a shape and size that is often characteristic of the larval species concerned.

Bressummer   A large timber beam found in older buildings, spanning a fireplace or a bay window. The built-in ends of this wooden member are highly prone to fungal attack.

Brown rot   Decay by wood-destroying fungi, in which the cellulose component of the wood cell walls is broken down. The wood develops cuboidal cracking, and is friable when handled.

Calcite   A crystalline form of calcium carbonate, often associated with igneous or sedimentary rocks.

Cambium   The zone in the trunk of a tree between the bark and the wood, where secondary thickening and growth in girth of the tree take place.

Capillarity   The surface tension forces that occur on the walls of capillaries, which may cause water or any liquid to flow into or be rejected from the surfaces.

Capillary   A thin, hollow tube. The extent to which water will rise in a capillary tube against the force of gravity is inversely proportional to the diameter of the capillary.

Capillary moisture (in a wall)   The moisture that rises in a wall as a result of rising capillarity in the fine cracks in the masonry, and especially in the mortar.

Cell walls   Developed from the plant cell wall membrane, they are composed of up to three layers of cellulose microfibrils orientated mainly along the grain of the wood, as a spiral, and embedded in a matrix of lignin. Wood properties, especially strength and movement, are much influenced by the cell wall structure and composition.

Cellulose   A polymeric constituent of plant cell walls, composed of glucose units, often forming microfibril structures orientated as spirals, generally in the direction of the length of the cell and the grain of the wood. They confer much of the tensile strength in the cell wall and thus in the wood.

Chipboard   A wood-based sheet material in which wood particles are bonded together using a resin-based glue in a heated press.

Chlorophyll   The green pigment in plant leaves, which is essential for the photosynthetic process.

Combined water   The water that is present in wood cell walls. Once the moisture content drops below fibre saturation point, the combined water dries, and some of it evaporates. This loss of water causes movement and reduction in electrical conductivity, but strength increases. If the moisture content increases again, the effects on these wood properties are reversed. Generally, it is the prevailing ambient humidity that determines the amount of combined water present. Thus wood moisture content changes if humidity conditions alter.

Condensation   The process of a liquid forming from its vapour. When moist air is cooled below the dew-point, water vapour will condense on cool surfaces as a liquid.

Cryptoclimate   The climate in and around a building and its elements (Chandler, 1989). Compare with macroclimate (the climate of a whole country of region); mesoclimate (the climate of a district or part of a country); microclimate (the climate of a site).

Crystal   A substance may become crystalline on drying or cooling, forming a structure in which all of the constituents are arranged in a regular three-dimensional pattern.

Crypto-florescence   Efflorescence forming at or near the surface of a material (e.g. limestone). This can cause exfoliation of the surface of the stone or brick (Addleson & Rice, 1991).

Dehydration   The removal of water by heating, chemical action or other means.

Density   Expressed as the mass or weight of a unit volume of a substance in kilograms per cubic metre (kg/m^3). Porous substances can have a wide range of densities depending on the volume of pores, e.g. wood and aerated concrete products. Generally, the highest densities are associated with the greatest strength (lead being an exception).

Desorption   This is a form of evaporation of moisture and is the reverse process to adsorption.

Dew-point   The temperature at which condensation occurs.

Diffusion   The general transport of matter whereby molecules or ions mix through normal thermal agitation. It is the movement of substances from a centre of high concentration to achieve a balanced concentration (equilibrium). This process is an example of the high–low principle (Addleson & Rice, 1991).

Dissolution   Soluble salts dissolving in water.

Dry rot   Decay in buildings caused by *Serpula lacrymans*. The name in one sense is apt, because the fungus can cause decay of wood at relatively low moisture content (down to 20%), although this is still wet in terms of normal equilibrium moisture contents of timbers in buildings.

Drying   The process of removal of water from a material. It can be achieved by air drying, kiln drying or by the use of a dehumidifier.

Efflorescence   The formation of a white crystalline deposit on the face of masonry due to the drying out of salts present in the units or the mortar.

Electro-osmosis   The movement of a liquid under an applied electric field through a fine tube or membrane.

Emulsion   The suspension of a substance in a liquid in which it cannot dissolve because it is immiscible.

Enzymes   Complex protein chemicals produced by living organisms that cause biochemical reactions to take place. They are catalysts produced by living cells. Wood-destroying fungi exude enzymes to digest wood cell walls as part of the decaying process.

Equilibrium moisture content   Many porous building materials have a moisture content that is normally in balance with the prevailing moisture conditions. The moisture content will be greater with materials of higher hygroscopicity.

Ettringite   A crystalline inorganic mineral formed in cement during the hydration of calcium aluminates in the presence of sulphate ions.

Eukaryote   Organism whose cells possess nuclei and other membrane-bounded organelles. Compare with prokaryote (Solomon *et al.*, 1996).

Evaporation   The conversion of a liquid into vapour, at a temperature below the boiling point. The rate of evaporation increases with rise of temperature, since it depends on the saturated vapour pressure of the liquid, which rises until it is equal to the atmospheric pressure at the boiling point.

Fibre   A thin wood cell up to 3 mm long with pointed ends, orientated along the grain and having a mechanical strength function.

Fibre saturation point   A moisture content of wood in the range 25–30%, when (theoretically) all the free water in the cell lumen has been lost as a result of drying but the combined water in the cell walls still remains.

Fungus   A heterotrophic eukaryote with chitinous cell walls and a body usually in the form of a mycelium of branched, thread-like hyphae, or unicellular (Solomon *et al.*, 1996). Most fungi are decomposers (i.e. saprophytes); some are parasitic. Fungi are nucleated organisms, which lack green colouring matter or chlorophyll, and therefore cannot photosynthesize their food (Singh, 1993). Basidiomycetes are the subgroup of fungi that are responsible for wood-rooting species causing dry rot and wet rot.

Grain   The direction of the axes of the principal cells in wood. Greatest tensile strength and permeability are to be found in the grain direction. Movement is greatest at right angles to the grain. The end of a piece of converted timber is called the end grain, and the side is the side grain. Permeability to liquids is greatest along the grain, penetrating from exposed end-grain areas.

Green timber   A log, a piece of lumber or freshly converted timber that has a high moisture content (50–200%) because it has not been dried.

Growth rings   Caused by alterations or the cessation in the growth of the cambium; often associated with seasonal growth patterns. Different forms of wood cells are produced, and the variations may be visible as a series of rings on the end grain.

Gymnosperms   A major group of plants with open seeds. This group includes all softwoods (i.e. needle-leafed trees). Compare with angiosperms.

Hardwood   Timber from a broad-leaf tree (an angiospermous plant).

Heartwood   The central zone of the trunk of a tree where active sap conduction and transpiration has ceased. On converted wood, heartwood is often darker coloured, and sometimes more resistant to decay and insect attack and less permeable to liquids and wood preservatives.

Heterotroph   Organism that cannot synthesize its own food from organic raw materials, and therefore must live either at the expense of other organisms or upon decaying matter (i.e. includes most bacteria and fungi). Compare with autotroph (i.e. organism that can synthetize organic compounds from raw materials – includes some bacteria). Heterotroph is also called consumer; autotroph is called producer (Solomon *et al.*, 1996).

Hydrolysis   Any irreversible chemical reaction initiated by or involving water.

Hydrophilic   A substance or surface that is attracted to water.

Hydrophobic   A substance or surface that is repellent of water. Compare with hydrophilic.

Hygrometer   An instrument for measuring atmospheric humidity. There are basically three types of hygrometer in use: the whirling hygrometer, the hair hygrometer, and the digital electronic hygrometer.

Hygroscopic   If a substance is hygroscopic, it has an attraction to water. Hygroscopic substances may absorb or lose water to the environment as they attempt to achieve a balance with the prevailing atmospheric relative humidity.

Hygroscopic moisture   The moisture in a material that can be attributed to the hygroscopic materials present. Thus for wood, the cellulose attracts atmospheric moisture; in masonry, hygroscopic salts act in a similar fashion. The amount of hygroscopic moisture will be related to the prevailing ambient relative humidity in the surrounding atmosphere.

Hyphae   Tubular branching root-like structures that are produced by a fungus. They are microscopic in size, and penetrate into the substance on which the fungus is growing. The hyphae of wood-destroying fungi produce the enzymes which digest and decay the wood substance.

Igneous rock   A rock deriving directly from volcanic activity.

Immiscible   Incapable of being mixed.

Insoluble   Incapable of being dissolved.

Insulation   Any means of confining a transmission phenomenon, e.g. of heat or sound, to obviate or minimise loss or damage.

Kiln drying   Drying of a material by artificial means under high heating conditions. Timber can be dried in a kiln in which, by controlling air flow, humidity and temperature, it is possible to dry wood to the low moisture levels necessary in centrally heated buildings. Such kiln-dried wood, when dried to the appropriate equilibrium moisture content, should not exhibit movement in use – unless, of course, moisture conditions alter again.

Lignin   A phenolic-type polymer substance present in wood cells that binds the fibres together.

Macropores   Fine or small cavities in a material, between groups of particles.

Metamorphic rock   A rock formed as a result of the action of heat and pressure on sedimentary or igneous rock.

Micropores   Minute cavities in a material – between particles.

Moisture content   The amount of water present in a substance, expressed as a percentage of its oven-dry weight. In timber, it can be measured reasonably accurately with an electrical moisture meter.

Moisture movement: Many building materials may show movement, especially shrinkage on initial drying-out of moisture. If the material has a high degree of hygroscopicity, swelling movement may occur if the material increases in moisture content.

Mould   A superficial growth of fungus, which occurs under conditions of high humidity. The minimum level for mould growth is frequently quoted as 70 RH (Garratt & Nowak, 1991).

Mycelium   A mass of visible hyphae produced by a fungus, which can be seen on the surface of timber or masonry.

Mycology   The study of fungi (from the Greek *mykes*, mushroom or fungus; *logos*, discourse).

Myxomycota   The group of wall-less fungi (e.g. slime moulds). Compare with eumycota (the true walled fungi, such as basidiomycetes).

Natural durability   The inherent resistance of certain timbers to fungal decay. The classification of natural durability is based on the results of ground contact graveyard-type tests.

Osmosis   This is diffusion of a solvent through a semi-permeable membrane into a more concentrated solution, tending to equalise the concentrations on both sides of the membrane.

Pargetting   Coating of cement mortar in flues or between the underside joints of slates or tiles.

Permeability   The rate of diffusion of a gas or a liquid under a pressure gradient through a porous material. The permeability of wood describes the ease or difficulty of the flow of liquids through the wood. Greatest permeability is found along the grain direction.

Photosynthesis   The process carried out in green leaves of plants whereby carbon dioxide and water are combined in the presence of sunlight to form glucose, which is the basis of all the other products needed by the plant.

Plaster   Internal coating of walls and ceilings.

Plasticiser   An additive or admixture to cementitious materials, which improves workability. Plasticisers allow a reduction in the water : cement ratio, and can provide higher strength, lower shrinkage and greater frost resistance.

Plywood   A composite sheet material composed of wood veneers bonded with adhesive. Marine quality and WPB plywood contain water-resistant glues. Usually, the board is made of veneers in which the grain is laid at right angles to that of the adjacent ply to provide redistributed strength properties and restriction of movement. Marine plywood contains durable veneers.

Pointing   The process of raking out old mortar from masonry, and refilling with a compatible mortar.

Polymerisation   The combination of several broadly similar molecules to form a more complex molecule with the same basic formula as the simple molecule.

Porosity   The percentage of space in a material.

Relative humidity   This is the existing water vapour pressure of the atmosphere, expressed as a percentage of the saturated water vapour pressure at the same temperature. RH is similar to the saturation percentage, but is a better measure of moisture in the atmosphere, because it indicates how near the quantity of water vapour present in the air is to saturation under the same conditions. The basic formula for RH is:

$$RH = \frac{\text{Actual water content}}{\text{Saturated water content}} \times \frac{100}{1}$$

Render   A coating applied to the external surface of masonry to improve appearance and weather resistance.

Retrofit   A modern engineering term imported from North America. To retrofit is to introduce a modern component into a redundant or outmoded subject to enhance, revitalise or, indeed, reinstate its original effectiveness.

Salt   A substance derived from the action of an acid on a metal. Salts are generally crystalline when dry, but form ions when dissolved in water.

Sapwood   The outer zone of wood in the tree where active conduction of sap occurs. In converted timber, the sapwood is usually light coloured, less resistant to decay, more susceptible to insect attack and, when dry, more permeable to wood preservatives.

Sarking   An internal lining in a pitched roof under tiles or slates.

Seismic staining   Uncontrolled flow of rainwater under window sills, string courses, parapets, edge beams etc., giving an irregular seismic pattern of staining on wall surfaces (Addleson, Vol. 2, 1971). Lack of adequate drips or discontinuity of drip to underside of sills etc. is one of the main causes of this problem.

Sick building syndrome (SBS)   A phenomenon in which some or many occupants of a property complain of ailments directly resulting from using the building. Not everybody who works in such a building will necessarily become sick. The main symptoms associated with SBS are: dryness of skin, eyes, nose and throat; allergic effects – watery eyes, runny nose; asthmatic effects – chest tightness, wheezing; and general ailments such as lethargy and headache.

Softwood   Timber from coniferous cone-bearing trees, which are usually evergreen and have needle-shaped leaves (a gymnospermous plant).

Soluble   A substance that dissolves in a liquid is soluble in that liquid.

Solvent   A liquid in which other substances are dissolved. The movement of the solvent may cause the dissolved substances to be carried to other places, where they may be deposited if the solvent evaporates.

Solvents are used to carry the active ingredients of wood preservatives or damp-proofing chemicals into wood or masonry. Penetrating damp can act as a solvent for mineral salts present in masonry.

Specific humidity   This is a measure of the mass of water vapour per unit mass of moist air. It is also known as the mixing ratio, and is expressed in kg/kg of dry air involved.

Spore   A reproductive fungal cell, microscopic in size, which is produced by the fungus and serves to distribute the species to a fresh substrate or to enable the fungus to survive unfavourable environmental conditions.

Sporophore   A spore-producing structure (i.e. a fruiting body) of a fungus.

Strands   Conducting structures produced by the mycelium of certain wood-destroying fungi that assist in the conduction of water and other essential substances to the growing fringe of the mycelium. Sometimes referred to as rhizomorphs.

Tanking   The application of an impervious material to a wall to prevent penetrating, usually in below-ground situations.

Tracheid   A fibre-shaped wood cell up to 3 mm long, orientated along the grain, that performs both mechanical and conductive functions. It constitutes the major wood tissue in softwoods.

Vapour check   A layer that checks water vapour movement, and so prevents or minimises water condensing in an insulant around a cold surface.

Water : cement ratio   The strength of concrete and other cementitious products depends on the ratio of the weight of water to the weight of cement: the lower the ratio, the higher the strength of the concrete. The minimum ratio for a workable mix is 0.45, but 0.25 is sufficient to set the cement.

Water-repellence   The ability of a substance to repel water. A liquid water repellent can be applied to walls to encourage drying by preventing further water uptake. Water repellents are usually hydrophobic.

Water vapour pressure   That part of atmospheric pressure due to water vapour.

Weathering   The gradual process by which materials on the external faces of a building are affected by natural climatic conditions. It can take many forms: general wetting of the fabric; seismic staining; bleaching of paint and pigmented finishes, etc.

Wet rot   The decay of timber by one of several different species of wood-destroying fungus that require a relatively high timber moisture content (minimum 30%). (Excludes dry rot, *Serpula lacrymans*.)

White rot   A category of fungus that can decay both the lignin and the cellulose of the wood cell wall. Characteristically, the wood is bleached, turns an off-white, and becomes soft and springy. Cuboidal cracking does not develop. All white rots are wet rots.

Woodworm   A popular term that describes all forms of insect damage to timber.

Bibliography

References

Addleson, L. (1992) *Building Failures: A Guide to Diagnosis, Remedy and Prevention*, 3rd edn. Butterworth-Heinemann, London.

Addleson, L. & Rice, C. (1991) *Performance of Materials in Building*. Butterworth-Heinemann, London.

Allen, P. (1972) Changing patterns of living in homes. In: *Condensation in Buildings* (eds D.J. Croombe and A.F.C. Sherratt), pp. 19–23. Applied Science Publishers Ltd, London.

Anderson, J.M. & Gill, J.R. (1988) *Rainscreen Cladding: A Guide to Principles and Practice*. CIRIA and Butterworths, London.

Andrew, C., Young, M., Tonge, K. & Urquart, D. (1994) *Stonecleaning: A Guide for Practitioners*. Historic Scotland and the Robert Gordon University, Edinburgh.

Anon (1986) Construction risks and remedies: timber decay. *Architects Journal*, 8 October, 57–65; 15 October, 69–83.

Anon (1987a) Construction risks and remedies: movement. *Architects Journal*, 21 January, 57–68; 28 January, 49–57.

Anon (1987b) Construction risks and remedies: condensation. *Architects Journal*, 9 April, 49–58; 16 April, 69–81.

Anon (1987c) Construction risks and remedies: indoor air quality. *Architects Journal*, 10 June, 55; 17 June, 57.

Anon (1987d) Construction risks and remedies: thermal insulation. *Architects Journal*, 11 June, 57–64; 19 June, 61–77.

Ashurst, J. & Ashurst, N. (1988) *Practical Building Conservation*, Vol. 2: *Brick, Terracotta and Earth*. English Heritage Technical Handbook, Gower, Aldershot.

Bech-Anderson, J. (1991) *The Dry Rot Fungus and Other Fungi in Houses*. Document IRG/WP/2389, Hussvamp Laboratories, Denmark.

Beech, J.C. (1981) *The Selection and Performance of Sealants*. IP 25/81, Building Research Establishment, Garston.

Berry, R.W. (1994) *Remedial Treatment of Wood Rot and Insect Attack in Buildings*. Building Research Establishment, Garston.

Botsai, E. (1991) Quoted in J. Wilson, *Evaluating Building Materials*. J. Wiley & Sons, New York.

Boynton, R.S. (1980) *Chemistry and Technology of Lime and Limestone*, 2nd edn. John Wiley & Sons, Chichester.

Brand, S. (1994) *How Buildings Learn*. Viking, New York.

Bravery, A.F. (1991) Strategies for the eradication of *Serpula lacrymans*. In: *Serpula lacrymans: Fundamental Biology and Control Strategies* (eds D.H. Jennings and A.F. Bravery), pp. 117–130. John Wiley & Sons Inc, New York.

Bravery, A.F. & Carey, J.K. (1993) *Recognising Wood Rot and Insect Damage in Buildings*, 2nd edn. Building Research Establishment, Garston.

BRE (1991) *A Practical Guide to Infra-red Thermography for Building Surveys*. BRE Report 176, Building Research Establishment, Garston.

BRE (1994) *Thermal insulation: avoiding risks. A good practice guide supporting Building Regulations Requirements*. BRE Report 143, Building Research Establishment, Garston.

BRE & BRESCU (1991) *NHBC Guide to Thermal Insulation and Ventilation*. Good Practice Guide, National House-Building Council, London.

Brick Development Association (1989). Technical Information Paper 7. *Bricks: Notes on their Properties*.

Briggs Amasco Ltd (1983) *Flat Roofing*. RIBA Publications, London.

British Gypsum (1991) *The White Book: The Compendium of Dry Lining Systems and Plastering Products*. British Gypsum, Loughborough.

Brundrett, G.W. (1990) *Criteria for Moisture Control*. Butterworths, London.

Bryant, R. (1979) *The Dampness Monster: A Report on the Gorbals Anti-Dampness Campaign*. Scottish Council of Social Service, Edinburgh, UK.

Cairns, A.H. (1992) *New Construction Materials in the Design, Maintenance and Refurbishment of Buildings*, Vol. 1. Heriot-Watt University, Edinburgh.

Carter, L. & Skipper, S. (1992) The roof that leaked . . . (Parts 1 and 2). *The Building Surveyor*, **1** (5), 10–11; **1** (6), 8–9.

Chandler, I. (1992) *Repair and Refurbishment of Modern Buildings*. Batsford, London.

Christian, J.E. (1993) A search for moisture sources. In: *Bugs, Mold and Rot II*, Proceedings of a Workshop on Control of Humidity for

Health, Artifacts and Buildings, pp. 71–81. National Institute of Building Sciences, Washington, DC.

Chudley, R. (1995) *Building Construction Handbook*, 2nd edn. Heinemann-Newnes, London.

CIRIA (1995a) *Water-resisting Basements*. Report 139, Construction Industry Research and Information Association, London.

CIRIA (1995b) *Water-resisting Basements (summary report)*. Report 140, Construction Industry Research and Information Association, London.

CIRIA (1995c) *Wall Technology*, Vol. A: *Performance Requirements*. Special Publication 87, Construction Industry Research and Information Association, London.

CIRIA & BFRC (1993) *Flat Roofing: Design and Good Practice*. CIRIA and BFRC, London.

Clifton-Taylor, A. (1962) *The Pattern of English Building*. B.T. Batsford, London.

Coggins, C.R. (1991) Growth characteristics in a building. In: *Serpula Lacrymans: Fundamental Biology and Control Strategies* (eds D.H. Jennings and A.F. Bravery), p. 81–86. John Wiley & Sons, Chichester.

Connolly, J.D. (1993) Humidity and building materials. In: *Bugs, Mold and Rot II*, Proceedings of a Workshop on Control of Humidity for Health, Artifacts and Buildings, pp. 29–36. National Institute of Building Sciences, Washington, DC.

Cook, G.K. & Hinks, A.J. (1992) *Appraising Building Defects*. Longman, London.

Coote, A.T. (1975) *Rising Damp: Remedial*. BRE Seminar Notes Ref: B475/75, Building Research Establishment, Garston.

Covington, A., Bravery, A.F. & Wynands, R.H. (1992) *Moisture Conditions in the Walls of Timber-Framed Housing*. BRE Report 228, Building Research Establishment, Garston.

Covington, A., McIntyre, J.S. & Stevens, A. (1995) *Timber Frame Housing 1920–1975: Inspection and Assessment*. BRE Report, Building Research Establishment, Garston.

CSTC (1985) *Methods for Treating Rising Damp in Walls*. Technical Information Note 162, Centre Scientifique et Technique de la Construction, Brussels, Belgium.

Davey, A., Heath, R., Hodges, D., Ketchin, M. & Milne, R. (1995) *The Care and Conservation of Georgian Houses*, 4th edn. Butterworth Architecture, London.

Department of the Environment (1993) *English House Condition Survey: 1991*. HMSO, London.

Douglas, J. & Singh, J. (1995) Investigating dry rot in buildings. *Building Research and Information*, **23** (6) 345–352.

Duell, J. & Lawson, F. (1983) *Damp Proof Course Detailing*, 2nd edn., The Architectural Press, London.

Eaton, R.C. & Hale, M.D.C. (1993) *Wood: Decay, Pests and Protection*. Chapman & Hall, London.

Endean, K.F. (1995) *Investigating Rainwater Penetration of Modern Buildings*. Gower Publishing Ltd, Aldershot.

Essex, T. (1993) Counting the cost. *Building* (Windows and doors supplement, 30 April).

Euroroof Ltd (1985) *Re-Roofing*. Euroroof Ltd, Northwich, Cheshire.

Fearn, J. (1985) *Thatch and Thatching*. Shire Publications, Aylesbury.

Flannigan, B. (1992) Approaches to assessment of the microbial flora in buildings. In: *IAQ '92 Environment for People*, pp. 139–145. American Society of Heating, Refrigeration and Air-conditioning Engineers Inc., Atlanta, GA.

Flannigan, B. & Morey, P.R. (1996) *ISIAQ Task Force: Guidelines for Control of Moisture Problems Affecting Indoor Air Quality*. International Society of Indoor Air Quality and Climate, London.

Freeman, I.L., Butler, R.N. & Hunt, J.H. (1983) *Timber-framed Housing – a Technical Appraisal*. BRE Report 41, Building Research Establishment, Garston.

Garratt, J. & Nowak, F. (1991) *Tackling condensation: A guide to the causes of, and remedies for, surface condensation and mould in traditional housing*. BRE Report 174, Building Research Establishment, Garston.

Geary, P.G. (1970) *Measurement of Moisture in Solids*. Sira Institute, Kent.

Gratwick, R.T. (1966) *Dampness in Buildings*, Vols 1 and II combined. Crosby Lockwood, London.

Hall, N. (1988) *Thatching: A Handbook*. Intermediate Technology Publications, London.

Hamilton, W.N., Kennedy, P., Kilpatrick, A.R.M., Matherson, A. C. & McLaughlin, R.K. (1996) *The Scottish Building Regulations Explained and Illustrated*, 2nd edn. BSP Books, London.

Handisyde, C.C. (1967) *Building Materials: Science and Practice*, 6th edn. Architectural Press, London.

HAPM Ltd (1992) *Component Life Guide*. E. & F.N. Spon, London.

Harris & Edgar (1995) *Cladding Fixing Technical Handbook*. Harris & Edgar.

Harris, D. J. (1995) Moisture beneath suspended timber floors. *Structural Survey*, **13** (3), 11–15.

Haverstock, H. (1988) Thermal insulation. Easiguide 7, *Building Design* (17 June).

Holdsworth, W. & Sealey, A. (1992) *Healthy Buildings: A Design Primer for a Living Environment*. Longman, Harlow.

Hollis, M. (1991) *Surveying Buildings*, 3rd edn. RICS Books, London.

House of Commons (1984a) *Dampness in Housing*, Vol. I, Report, Proceedings of the Committee and Index, First Report from the Scottish Affairs Committee, Session 1983–1984, The House of Commons. HMSO, London.

House of Commons (1984b) *Dampness in Housing*, Vol. II, Minutes of Evidence and Appendices, First Report from the Scottish Affairs Committee, Session 1983–1984, The House of Commons. HMSO, London.

Houston, A.G. (1990) *Optimisms in chemically injected hydrophobic treatments in arenaceous media*. PhD Thesis, Heriot-Watt University, Edinburgh.

Howell, J. (1994) Diagnosis of rising damp and specification of remedial damp-proofing treatments. In: *RICS Focus for Building Surveying Research*, pp.15–20. The Royal Institution of Chartered Surveyors, London.

Howell, J. (1995) Moisture measurement in masonry: guidance for surveyors. In: *COBRA '95*, RICS Construction and Building Research Conference, Heriot-Watt University, 8–9 September, pp.145–149. The Royal Institution of Chartered Surveyors, London.

Howell, J. (1996) Quoted by T. Rowland in: Rising damp – the dry facts and fiction. *The Telegraph*, Property section (Wednesday 22 May), London.

Hunt, S. (1989) Effects of damp housing on health. *Health and Hygiene*, **10**, 12–15.

Hutton, T.C., Lloyd, H. & Singh, J. (1991) The environmental control of timber decay. *Structural Survey*, **10** (1), pp. 5–20. Henry Stewart Publications, London.

IMBM (1986) *Condensation*. Technical Paper No. 1, February 1986, The Institute of Maintenance and Building Management, Farnham, Surrey.

Kubal, M.T. (1993) *Waterproofing the Building Envelope*. McGraw-Hill, New York.

Kyte, C.T. (1984) *Laboratory Analysis as an Aid to the Diagnosis of Rising Damp*. Technical Information Note 35, Chartered Institute of Building, Ascot.

Kyte, C.T. (1987a) The treatment of rising damp. *Structural Survey*, **6** (3), 312–315.

Kyte, C.T. (1987b) *Laboratory analysis as an aid to the diagnosis of rising damp*. Technical Information Paper 35, Chartered Institute of Building, Ascot.

Lacey, J. (1994) Indoor aerobiology and health. In: *Building Mycology* (ed. J. Singh), pp. 77–129. E. & F.N. Spon, London.

Lee, F.M. (1980) *The Chemistry of Cement and Concrete*. Edward Arnold, London.

Lim, W.B.P. (ed.) (1988) *Control of the External Environment of Buildings*. Singapore University Press, Singapore.

London Hazards Centre (1990) *Sick Building Syndrome: Causes, Effects and Control*. London Hazards Centre Trust Ltd, London.

Markus, T. (1993) Cold, condensation and housing poverty. In: *Unhealthy Housing: Research Remedies and Reforms* (eds R. Burridge and D. Ormandy). E. & F.N. Spon, London.

Marley (1983) *Damp Proof Detailing*. Marley Waterproofing Ltd, Sevenoaks, Kent.

Marsh, P. (1977) *Air and Rain Penetration of Buildings*. Construction Press, London.

McCann, J. (1983) *Clay and Cob Buildings*. Shire Publications Ltd.

Melville, I.A. & Gordon, I.A. (1973) *The Repair and Maintenance of Houses*. Estates Gazette, London.

NBA (1985) *Maintenance Cycles and Life Expectancies of Building Components and Materials: A Guide to Data and Sources*. National Building Agency, London.

Neville, A.M. & Brooks, J.J. (1987) *Concrete Technology*. Longman Scientific & Technical, Harlow.

Newman, A.J. (1988) *Rain penetration through masonry walls: diagnosis and remedial measures*. BRE Report BR 117, Building Research Establishment, Garston.

NHBC (1993) *Technical Standards*, Vols 1–2. National House-Building Council, London.

Oliver, A.C. (1985) *Woodworm, Dry Rot and Rising Damp*. Sovereign Chemical Industries Ltd, Barrow-in-Furness, Cumbria.

Oxley, T.A. & Gobert, E.G. (1994) *Dampness in Buildings: Diagnosis, Treatment, Instruments*, 2nd edn. Butterworth-Heinemann, Oxford.

Phillipson, M.C. (1966) *Effects of moisture on porous masonry: overview of literature*. BRE Report 304, Building Research Establishment, Garston.

Platt, S., Martin, C. & Hunt, S. (1989) Damp housing, mould growth and symptomatic health state. *British Medical Journal*, **298**, 1673–1678.

Pountney, M.T., Maxwell, R. & Butler, A.J. (1988) *Rain Penetration of*

Cavity Walls: A Report of a Survey of Properties in England and Wales.
IP 2/88, Building Research Establishment, Garston.

Powell-Smith, V. & Billington, M.J. (1995) *The Building Regulations Explained and Illustrated*, 10th edn. Collins, Oxford.

Property Services Agency (1989) *Defects in Buildings*. HMSO, London.

Ragsdale, L.A. & Raynham, E.A. (eds) (1972) *Building Materials Technology*. Edward Arnold, London.

Ransom, W.H. (1987) *Building Failures: Diagnosis and Avoidance*, 2nd edn. E. & F.N. Spon, London.

Richardson, B.A. (1991) *Defects and Deterioration in Buildings*. E. & F.N. Spon, London.

Richardson, B.A. (1993) *Wood Preservation*. E. & F.N. Spon, London.

Rickards, M. (1987) If you can't see it, it probably isn't rising damp. *Structural Survey*, **5** (1), 233–236.

RICS Building Conservation Group (1993) *A Checklist for the Structural Survey of Period Timber Framed Buildings*, RICS Books, London.

RICS Building Surveyors' Division (1979) *Condensation Problems: New and Existing Buildings*. Practice Note No. 2, The Royal Institution of Chartered Surveyors, London.

Ridley, P. (1988) *Evaluation and Repair*. The Birmingham TERN Project, Birmingham Polytechnic.

Ridout, B.V. (1992) Dry rot: its history and treatment. *The Building Surveyor*, **1** (6), 6–7.

Ridout, B.V. (1994) *Timber in Buildings: Decay, Treatment and Preservation*. E. & F.N. Spon, London.

Rixom, M.R. (1977) *Concrete Admixtures*. Construction Press, Lancaster.

Rose, W.B. (1994) A review of the regulatory and technical literature related to crawl space moisture control. *ASHRAE Transactions*, **100** Pt 1.

Ross, K.D. & Butlin, R.N. (1989) *Durability Tests for Building Stone*. BRE Report 141, Building Research Establishment, Garston.

Rowland, T. (1996) Rising damp – the dry facts and fiction. *The Telegraph*, Property section (Wednesday 22 May), London.

Rushton, T. (1992) Understanding why buildings fail and using 'HEIR' Methodology in Practice. In: *Proceedings of Conference on Latest Thinking on Building Defects in Commercial Buildings: Their Diagnosis, Causes and Consequences*, 15 July 1992, pp. 3–15. Henry Stewart Conference Studies, London.

Scaduto, J.V. (1989) *What's It Worth: A Home Inspection and Appraisal Manual*, 2nd edn. TAB Books, Summit, PA.

Scottish Homes (1993) *Scottish House Condition Survey 1991: Key Findings.* Scottish Homes, Edinburgh.

Scottish Lime Centre (1995) *Preparation and Use of Lime Mortars: An Introduction to the Principles of Using Lime Mortars.* Technical Advice Note 1, Historic Scotland, Edinburgh.

Seiffert, K. (translated by Phillips and Turner) (1970) *Damp Diffusion and Buildings: Prevention of Damp Diffusion Damage in Building Design.* Elsevier, London.

Singh, J. (1993) Biological contaminants in the built environment and their health implications. *Building Research and Information,* **21** (4), 216–223.

Singh, J. (1994) The built environment and the development of fungi. In: *Building Mycology* (ed. J. Singh), pp. 1–21. E. & F.N. Spon, London.

Solomon, E.P., Berg, L.R., Martin, D.W. & Villee, C. (1966) *Biology,* 6th edn. Saunders College Publishing, New York.

Son, L.H. & Yuen, G.C.S. (1993) *Building Maintenance Technology.* Macmillan, London.

Strachan, D. (1993) Dampness, mould growth and respiratory disease in children. In: *Unhealthy Housing: Research Remedies and Reforms* (eds R. Burridge and D. Ormandy), E. F.N. Spon, London.

Sundell, J. (1994) On the association between building ventilation characteristics, some indoor environmental exposures, some allergic manifestations and subjective symptom reports. *Indoor Air,* International Journal of Indoor Air Quality and Climate, Supplement No. 2/94 (May), pp. 11–49. Institute of Environmental Medicine, Stockholm.

Taylor, G.D. (1992) *Materials of Construction,* 2nd edn. Longman, Harlow.

Thomson, J.C. (1994) Frozen waste. *Building,* **CCLIX** (23 September), 52.

Tovey, A.K. & Roberts, J.J. (1990) *Efficient Masonry Housebuilding: Detailing Approach.* British Cement Association, Slough.

Trotman, P. (1995) Dampness in cob walls. In: *Out of Earth II,* p. 118, Plymouth University, Plymouth.

Verhoef, L.G.W. (1988) *Soiling and Cleaning of Building Facades.* RILEM Report, Chapman & Hall, London.

Whitely, P., Russman, H.D. & Bishop, J.D. (1977) Porosity of building materials. *Journal of the Oil and Colour Chemists' Association,* **60,** 142–150.

Wingate, M. (1990) *An Introduction to Building Limes.* Information Sheet 9, Society for the Protection of Ancient Buildings, London.

Wright, A. (1991) *Craft Techniques for Traditional Buildings.* B.T. Batsford, London.

BRE publications

The BRE produces a wide range of very informative publications on all aspects of building technology other than the two main groups listed below. It publishes Information Papers, Overseas Building Notes, Reports, and miscellaneous leaflets. The following Digests and Good Building Guides are relevant to dampness.

Digests

Digest 27 (1969) Rising damp in walls (withdrawn)
Digest 54 (1971) Damp-proofing solid floors
Digest 108 (1991) Standards (U-values)
Digest 110 (1972) Condensation
Digest 125 (1988) Colourless treatment of masonry
Digest 144 (1972) Asphalt and built-up felt roofing
Digest 145 (1972) Heat losses through ground floors
Digest 152 (1973) Repair and renovation of flood-damaged buildings
Digest 157 (1992) Calcium silicate brickwork
Digest 163 (1974) Drying out buildings
Digest 170 (1984) Ventilation of internal bathrooms and WCs in dwellings
Digest 177 (1975) Decay and conservation of stone masonry
Digest 180 (1986) Condensation in roofs
Digest 190 (1976) Heat losses from dwellings
Digest 196 (1976) External rendered finishes
Digest 197 (1982) Painting walls: Part 1 – choice of paint
Digest 198 (1977) Painting walls: Part 2 – failures and remedies
Digest 217 (1978) Wall cladding defects and their diagnosis
Digests 227, 228 and 229 (1979) Estimation of thermal and moisture movements and stresses (Parts 1, 2 and 3)
Digest 236 (1980) Cavity insulation
Digest 238 (1992) Reducing the risk of pest infestation: design recommendations and literature review
Digest 244 (1984) Rising damp in walls – diagnosis and treatment
Digest 269 (1983) The selection of natural building stone
Digest 270 (1983) Condensation in insulated domestic roofs
Digest 280 (1983) Cleaning external surfaces of buildings
Digest 296 (1985) Timbers – their natural durability and resistance to preservative treatments
Digest 297 (1985) Surface condensation and mould growth in traditional dwellings

Digest 299 (1985) Dry rot: its recognition and control
Digest 301 (1985) Corrosion of metals by wood
Digest 304 (1985) Preventing decay in external joinery
Digest 306 (1986) Domestic draughtproofing: ventilation considerations
Digest 307 (1986) Identifying damage by wood-boring insects
Digest 312 (1986) Flat roof design: the technical options
Digest 321 (1987) Timber for joinery
Digest 324 (1987) Flat roof design: insulation
Digest 327 (1992) Insecticidal treatments against wood-boring insects
Digest 329 (1993) Installation of wall ties in existing construction
Digest 330 (1988) Alkali aggregate reactions in concrete
Digest 336 (1988) Swimming pool roofs: minimizing the risks of condensation
Digest 340 (1989) Choosing wood adhesives
Digest 345 (1989) Wet rots: recognition and control
Digest 350 (1990) Climate and site development
Digest 351 (1990) Recovering old timber roofs
Digest 354 (1990) Painting of exterior wood
Digest 362 (1991) Building mortar
Digest 363 (1991) Sulphate and acid resistance of concrete in the ground
Digest 364 (1991) Design of timber floors to prevent decay
Digest 365 (1991) Soakaway design
Digest 369 (1992) Interstitial condensation and fabric deterioration
Digest 370 (1992) Control of lichens, moulds and similar growths
Digest 371 (1992) Remedial wood preservatives: use them safely
Digest 372 (1992) Flat roof design: waterproof membranes
Digest 373 (1992) Wood chipboard
Digest 375 (1992) Wood-based products: their contribution to the conservation of forest resources
Digest 378 (1993) Wood preservatives: application methods
Digest 380 (1993) Damp-proof courses
Digest 387 (1993) Natural finishes for exterior wood
Digest 392 (1993) Assessment of existing high alumina cement construction in the UK
Digest 405 (1993) Carbonation of concrete and its effects on durability

Defect Action Sheets

This excellent series of concise but informative articles, which ran from May 1982 to March 1990, has been superseded by the Good Building Guide series. Much of the material in the Defect Action Sheets, however, has been incorporated in the PSA book *Defects in Buildings* as well as in some of the following.

Good Building Guides

Good Building Guide 3 (1990) Damp proofing existing basements

Good Building Guide 5 (1990) Choosing between cavity, internal and external wall insulation

Good Building Guide 6 (1990) Outline guide to assessment of traditional housing for rehabilitation

Good Building Guide 9 (1991) Habitability guidelines for existing housing

Good Building Guide 11 (1993) Supplementary guidance for assessment of timber framed houses, Part 1 Examination

Good Building Guide 12 (1993) Supplementary guidance for assessment of timber framed houses, Part 2 Interpretation

Good Building Guide 22 (1995) Maintaining exterior wood finishes

Good Building Guide 23 (1995) Assessing external rendering for replacement or repair

Good Building Guide 24 (1993) Repairing external rendering

British Standards

BS 12 (1989) Portland cements

CP 102 (1973) Protection of buildings against water from the ground moisture

BS 144 (1990) Wood preservation by means of pressure creosote

CP 144 (1970) Mastic asphalt (Part 4)

BS 402 (1990) Plain tiles and fittings

BS 417 (1987) Galvanised mild steel cisterns and covers, tanks and cylinders

BS 437 (1978) Cast iron spigot and socket drainpipes and fittings

BS 680 (1971) Roofing slates (Part 2)

BS 743 (1970) Materials for damp proof courses

BS 747 (1994) Roofing felts

BS 776 (1972) Specification for materials for magnesium oxychloride (Magnesite) flooring

BS 882 (1992) Aggregates from natural sources for concrete

BS 890 (1995) Building limes

BS 1105 (1994) Wood wool cement slabs up to 125 mm thick

BS 1142 (1989) Fibre building boards

BS 1178 (1982) Milled sheet lead for building purposes

BS 1191 Part 1 (1994) Gypsum building plasters

BS 1191 Part 2 (1991) Premixed lightweight plasters

BS 1196 (1989) Clayware field drain pipes and junctions

BSs 1199–1200 (1984/86) Building sands from natural sources

BS 1230 (1994) Gypsum plasterboard

BS 1243 (1978) Metal ties for cavity wall construction

BS 2592 (1973) Specification for thermoplastic flooring tiles

BS 2870 (1980) Rolled copper and copper alloys

BS 3260 (1991) Semi-flexible floor tiles

BS 3261 (1973) Specification for unbonded flexible PVC floor coverings

BS 3505 (1986) Unplasticised PVC pipe for cold potable water

BS 3921 (1995) Specification for clay bricks

BS 4016 (1972) Building papers (breather type).

BS 4072 (1987) Wood preservative by means of copper/chrome/arsenic compositions

BS 4514 (1983) Upvc soil and ventilating pipes, fittings and accessories

BS 4551 (1980) Methods of testing mortar screeds and plasters

BS 4576 (1989) Upvc rainwater goods and accessories (Part 1)

BS 4660 (1989) Upvc pipes and plastic fittings for below ground drainage and sewers

BS 4756 (1991) Ready-mixed aluminium priming paints for woodwork

BS 5082 (1993) Water-borne priming paints for wood

BS 5224 (1993) Masonry cement

BS 5250 (1995) Control of condensation in buildings

BS 5262 (1991) External rendered finishes

BS 5268 (1989) Structural use of timber

BS 5390 (1977) Stone masonry

BS 5492 (1990) Internal plastering

BS 5528 (1985) Use of masonry

BS 5534 (1990) Slating and tiling (Part 1: Design)

BS 5589 (1989) Preservation of timber

BS 5618 (1985) Thermal insulation of cavity walls by filling with urea-formaldehyde foam systems

BS 5624 (1983) Sills and copings

BS 5628 Part 3 (1985) Structural use of masonry

BS 5669 (1989) Specification for wood, chipboard, and methods of test for particle boards

BS 5707 (1979/1990) Parts 1, 2, 3: Wood preservatives in organic solvents

BS 6093 (1981) Design of joints and jointing in buildings

BS 6150 (1991) Painting of buildings

BS 6213 (1982) Selection of construction sealants

BS 6229 (1982) Flat roofs with continuously supported coverings

BS 6297 (1993) Design and installation of small sewage treatment works and cesspools

BS 6367 (1983) Drainage of roofs and paved areas

BS 6398 (1983) Bitumen damp-proof courses for masonry

BS 6375 (1983) Performance of windows

BS 6477 (1992) Water repellents for masonry surfaces

BS 6515 (1984) Polyethylene damp-proof courses for masonry

BS 6524 (1985) Code of practice for installation of damp-proof courses

BS 6566 (1988) Plywood

BS 6576 (1985) Installation of chemical damp-proof courses

BS 6577 (1985) Mastic asphalt for building (natural rock asphalt aggregate)

BS 7543 (1992) Guide to the durability of buildings and building elements, products and components

BS 8000 (1990) Workmanship of building sites (in 12 parts)

BS 8023 (1987) Sheet and tile flooring

BS 8100 (1985) Design of non-loadbearing external vertical enclosures of buildings

BS 8102 (1990) Protection of structures against water from the ground

BS 8203 (1987) Code of practice for the installation of sheet and tile flooring

BS 8204 (1987) Concrete bases and screeds to receive insitu floorings (Part 1)

BS 8208 (1985) Assessment of suitability of external cavity walls for filling with thermal insulation

BS 8210 (1982) Code of practice for flooring of timber, timber products and wood-based panel products

BS 8213 (1990) Installation of replacement windows and door sets in dwellings

BS 8215 (1991) Design and installation of damp-proof courses

BS 8217 (1994) Built up felt roofing

BS 8300 (1985) Building drainage

BBA (British Board of Agrément) Certificates

Agrément Certificate No. 95/3123 (1995): Wykamol Chemical Damp-Proofing System

Agrément Certificate No. 94/3068 (1994): Anaplast Visqueen

Agrément Certificate No. 93/2870 (1993): Stanhope Injection Mortar

Agrément Certificate No. 92/2849 (1992): Freezteq DPC system

Agrément Certificate No. 89/2299 (1989): Permaphalt
Agrément Certificate No. 85/1567 (1985): Isofoam CRF Cavity Wall
Stabilisation

BWPDA (British Wood Preserving and Damp-Proofing Association) publications

Timber leaflets

T1 (1995) Fungal decay in buildings – dry rot and wet rot
T2 (1993) Preservative treatment of timber
T3 (1993) Methods of applying preservatives
T4 (1993) The preservation of window joinery
T5 (1993) Preservative treatment against wood borers
T6 (1990) The use of creosote oil for wood preservation
T7 (1993) The treatment of timber with water borne preservatives
T8 (1993) Water repellent wood preservatives
T9 (1990) Preserving wood in home and garden
FR1 (1995) The treatment of solid timber and panel products with flame
retarders

Damp-proofing literature

DP 1 (1990) The use of moisture meters to establish the presence of rising
damp
DP 2 (1990) Plastering in association with damp-proof coursing
DP 3 (1990) Condensation
DP 4 (1990) Methods of analysis for damp-proof course fluids
DP 5 (1990) Chemical damp-proof course insertion – the attendant
problems
DP 6 (1990) Safety in damp-proofing
DP 7 (1993) Chemical damp-proof course in walls – detection techniques
and their limitations
DP 8 (1993) Damp-proof barriers (tanking) in association with chem-
nical damp-proof coursing
DP 9 (1993) Guidelines to survey report writing
DP 10 (1990) Hygroscopic salts and rising damp

Codes of Practice

Code of Practice for Remedial Timber Treatment (1990)

Code of Practice for the Installation of Chemical Damp-proof Courses (1986)
Code of Practice for Safe Design and Operation of Timber (1993)

Further reading

Addleson, L. (1971–1972) *Materials for Buildings*, Vols 1, 2 and 3. Iliffe Books, London.

Ashurst, J. & Ashurst, N. (1988) *Practical Building Conservation*, Vols 1–5. English Heritage Technical Handbook, Gower Technical Press, Aldershot.

Billington, N.S. (1967) *Building Physics: Heat.* Pergamon Press, London.

Bonshor, R.B. & Bonshor, L.L. (1996) *Cracking in Buildings*, Construction Research Communications Ltd.

Bowen, R. (1982) *Surface Water.* Chapman & Hall, London.

Bowen, R. (1986) *Groundwater*, 2nd edn. Chapman & Hall, London.

Anon. (1959 and 1962) *Principles of Modern Building*, Vols 1 and 2. HMSO, London.

Burridge, R. & Ormandy, D. (eds) (1995) *Unhealthy Housing: Research, Remedies and Reform.* E.&F.N. Spon, London.

Cairns, A.H. (1992) *New Construction Materials in the Design, Maintenance and Refurbishment of Buildings*, Vols 1 and 2. Heriot-Watt University, Edinburgh.

Chandler, I. (1989) *Building Technology 2 – Performance.* Mitchell/CIOB, London.

Cook, G.K. & Hinks, A.J. (1992) *Appraising Building Defects.* Longman, London.

Curwell, S.R. & March, C.G. (eds) (1986) *Hazardous Building Materials: A Guide to the Selection of Alternatives.* E.&F.N. Spon, London.

HAPM Ltd (1991) *Defects Avoidance Manual: New Build.* Building Research Establishment, Garston.

Illston, J.M. (ed.) (1993) *Construction Materials – Their Nature and Behaviour.* Chapman & Hall, London.

Ineichen, B. (1993) *Homes and Health.* Chapman & Hall, London.

Marley (1989) *The damp proofing design guide.* Marley Waterproofing Ltd.

Matulionis, R.C. & Freitag, J.C. (eds) (1990) *Preventive Maintenance of Buildings.* Van Nostrand Reinold, New York.

McCambell, B.H. (1991) *Problems in Roofing Design.* Butterworth Architecture, London.

Melville, I.A. & Gordon, I.A. (1992) *Structural Surveys of Dwelling Houses*, 3rd edn. Estates Gazette, London.

Mika, S.L.J. & Desch, S.C. (1988) *Structural Surveying*, 2nd edn. Macmillan, London.

Noy, E.A. (1995) *Building Surveys and Reports*, 2nd edn. Blackwell Science, Oxford.

Parnham, P. (1996) *Prevention of Premature Staining of New Buildings*. Chapman & Hall, London.

Pavey, N. (1995) *Rules of Thumb*. Technical Note TN 17/95 (2nd edn). Building Services Research and Information Association, Bracknell, Berkshire.

Price, M. (1996) *Introducing Groundwater*, 2nd edn. E.&F.N. Spon, London.

Ranson, R. (1991) *Healthy Housing: A Practical Guide*. Chapman & Hall, London.

Richardson, B.A. (1995) *Remedial Treatment of Buildings*, 2nd edn. Butterworth.

Seeley, I.H. (1987) *Building Maintenance*, 2nd edn. Macmillan, London.

Singh, J. (ed.) (1994) *Building Mycology: Management of Decay and Health in Buildings*. Chapman & Hall, London.

Solomon, E.P., Berg, L.R., Martin, D.W. & Villee, C. (1996) *Biology*, 6th edn. Saunders College Publishing, New York.

Staveley, H.S. & Glover, P. (1990) *Building Surveys*, 2nd edn. Butterworth-Heinemann, London.

Universities of Sussex and Westminster (1996) *The Real Cost of Poor Homes*. The Royal Institution of Chartered Surveyors, London.

Woolman, R. (1994) *Resealing of Buildings: A Guide to Good Practice* (ed. by A. Hutchinson). Butterworth-Heinemann, Oxford.

Index